AF531350

ENCYCLOPAEDIA OF STEM CELLS - IV

ANIMAL STEM CELLS

By

Dr. Amita Sarkar

Dept. of Zoology

Agra College

Agra (U.P.)

(India)

First Published – 2008

Reprinted – 2025

ISBN: 978-81-8356-358-1 (Set)
978-81-8356-402-1

Animal Stem Cells

Published by:

DISCOVERY PUBLISHING HOUSE
4383/4B, Ansari Road, Darya Ganj
New Delhi-110 002 (India)
Phone: +91-11-23279245; 23253475; 43596065
Mobile: +91 9811179893 / +91 9871656464
E-mail: discoverybooksindia@gmail.com
orderdphbooks@gmail.com
namitwasan9@gmail.com
web: www.discoverypublishinggroup.com

Printed at:
Infinity Imaging Systems
Delhi

Preface

The present title *Encyclopaedia of Stem Cells* is the amazing advancement of biotechnology. It provides the various fundamental aspects of stem cell technologies to be understood adequately. It has been compiled for graduate and undergraduate students, research scholars, teachers, practising biochemical engineers, biotechnologists, applied and industrial microbiologists, cell biologists and scientists involved in bioprocessing research and development. The basic concepts have been clearly explained and their functions are adequately highlighted. The presentation of the text is simple and systematic. The selection of chapters and topics is according to the specified syllabi of several Indian Universities and are so structured as to enable the student to move easily from the fundamental to the complex. It is our earnest hope that this title will be of great value to all our students.

To make the work more comprehensive and informative, the author has consulted many authoritative books, research journals, abstracts, monographs etc. He is grateful to all those great scholars whose work are cited or substantially reproduced.

There can be no claim to originality except in the manner of treatment and much of the information has been obtained from the books and scientific journals available in the different libraries.

The author expresses his thanks to his friends and colleagues whose continue inspirations have initiated him to bring out this book.

The author expresses his gratitude to Mr. Wasan and staff of M/s Discovery Publishing Pvt. Ltd. for their whole hearted co-operation in the publication of this book.

In the mean time, the author will remain sincerely responsible for any shortcomings of the book and the grateful to the readers for their suggestions and constructive criticism for the continuous betterment of the book. He takes this opportunity to appeal to the readers to send their suggestions straightway to his publisher.

Author

CONTENTS

1

INTRODUCTION

Cell culture has become an indispensable technology in many branches of the life sciences. It provides the basis for studying the regulation of cell proliferation, differentiation and product formation in carefully controlled conditions, with processes and analytical tools which are scalable from the level of the single cell to in excess of 10 kg wet weight of cells. Cell culture has also provided the means to define almost the entire human genome, and to dissect the pathways of intracellular and intercellular signalling which ultimately regulate gene expression. From its ancestry in developmental biology and pathology, this discipline has now emerged as a tool for molecular geneticists, immunologists, surgeons, bioengineers, and manufacturers of Pharmaceuticals, while still remaining a fundamental tool to the cell biologist, whose input is vital for the continuing development of the technology.

Cell culture has matured from a simple, microscope driven, observational science to a universally acknowledged technology with roots set as deep in industry as they are in academia. It stands among microelectronics, avionics, astrophysics and nuclear engineering, as one of the major bridges between fundamental research and industrial exploitation, and, in the current climate, perhaps the more commercial aspects will ensure its development for at least several more decades. The prospects for genetic therapy and tissue replacement are such that the questions are rapidly becoming ethical, as much as technical, as new opportunities arise for genetic manipulation, whole animal cloning and tissue transplantation. Several textbooks are already available to introduce the complete novice to the basic principles of preparation, sterilization, and cell propagation, so this book will concentrate on certain specialized aspects, many of which are essential for the complete understanding and correct application of the technology. This chapter will review some of the biology of cultured cells, their derivation and characterization, and give guidance in those areas which require critical observation and control.

Biology of Cells in Culture

Origin and Characterization

The list of different cell types which can be grown in culture is extensive, includes representatives of most major cell types, and has significantly increased since the last edition of this book, due largely to the improved availability of selective media and specialized cell cultures through commercial sources such as Clonetics.

The use of markers that are cell type specific has made it possible to determine the lineage from which many of these cultures were derived, although the position of the cells within the lineage is not always clear. During propagation, a precursor cell type will tend to predominate, rather than a differentiated cell. Consequently a cell line may appear to be heterogeneous, as some cultures, such as epidermal keratinocytes, can contain stem cells, precursor cells, and mature differentiated cells. There is constant renewal from the stem cells, proliferation and maturation in the precursor compartment, and terminal and irreversible differentiation in the mature compartment. Other cultures, such as dermal fibroblasts, contain a relatively uniform population of proliferating cells at low cell densities (about 10^4 cells/cm^2) and an equally uniform, more differentiated, non-proliferating population at high cell densities (10^5 cells/cm^2). This high density population of fibrocyte-like cells can re-enter the cell cycle if the cells are trypsinized or scraped (or 'wounded' by making a cut in the monolayer) to reduce the cell density or create a free edge. Most of the cells appear to be capable of proliferation, and there is little evidence of renewal from a stem cell compartment.

Culture heterogeneity also results from multiple lineages being present in the cell line. The only unifying factors are the selective conditions of the medium and substrate, and the predominance of the cell type (or types) which have the ability to survive and proliferate. This tends to select a common phenotype but, due to the interactive nature of growth control, may obscure the fact that the population contains several distinct phenotypes only detectable by cloning.

Nutritional factors like serum, Ca^{2+} ions, hormones, cell and matrix interactions, and culture density, can affect differentiation and cell proliferation. Hence, it is not only essential to define the lineage of cells being used, but also to characterize and stabilize the stage of differentiation, by controlling cell density and the nutritional and hormonal environment to obtain a uniform population of cells, which will respond in a reproducible fashion to given signals.

As the dynamic properties of cell culture (proliferation, migration, nutrient utilization, product secretion) are sometimes difficult to control, and the complexity of cell interactions found *in vivo* can be difficult to recreate *in vitro*, there have been numerous attempts to either retain the structural integrity of the original tissue, using traditional organ culture, or to recreate it by combining propagated cells of different lineages in organotypic culture. Among the most successful examples of the latter are the so-called skin equivalent models, where epidermal keratinocytes are co-cultured with dermal fibroblasts and collagen in filter well inserts or some similar mechanical support. The result is a synthetic skin suitable for grafting and now being evaluated for tests of irritancy and inflammation.

Differentiation

As propagation of cell lines requires that the cell number increases, culture conditions have evolved to favour maximal cell proliferation. It is not surprising that these conditions are often not conducive to cell differentiation where cell growth is severely limited or completely abolished. Those conditions which favour cell proliferation are low cell density, low Ca^{2+} concentration (100-600 mM), and the presence of growth factors such as *epidermal growth factor* (EGF), *fibroblast growth factor* (FGF), and *platelet-derived growth factor* (PDGF). High cell density ($> 1 \times 10^5$ cells/cm^2), high Ca^{2+} concentration (300-1500 mM), and the presence of differentiation inducers, hormones such as hydrocortisone, paracrine factors such as IL-6 and KGF, and nerve growth factor, retinoids, and planar polar compounds, such as dimethyl sulfoxide, will favour cytostasis and differentiation.

The role of serum in differentiation is complex and depends on the cell type and medium used. While a low serum concentration promotes differentiation in oligodendrocytes, a high serum concentration causes squamous differentiation in bronchial epithelium. In the latter case, this is due to transforming growth factor β (TGFβ) released from platelets. Because of the undefined composition of serum and the ever-present risk of adventitious agents such as viruses, controlled studies on selective growth and differentiation are best conducted in serum-free media.

The establishment of the correct polarity and cell shape may also be important, particularly in epithelium. Many workers have shown that growing cells to high density on a floating collagen gel allows matrix interaction, access to medium on both sides and, particularly, to the basal surface where receptors and nutrient transporters are expressed. The plasticity of the substrate can facilitate a normal cell shape and normal polarity of the interactive ligands in the basement membrane.

Different conditions may be required for propagation and differentiation and hence an experimental protocol may require a growth phase for expansion and to allow for replicate samples, followed by a non-growth maturation phase to allow for increased expression of differentiated functions.

Choice of Materials

Cell Type

The cell type chosen will depend on the question being asked. For some processes, such as DNA synthesis, response to cytotoxins, or apoptosis, the cell type may not matter, provided the cells are competent. In other cases a specific process will require a particular cell type, for example surfactant synthesis in the lung will require a fresh isolate of type II pneumocytes or a cell line, such as NCI-H441, which still expresses surfactant proteins. A reasonable first step would be to determine from the literature whether a cell line exists with the required properties.

Source of Tissue

Embryo or adult?

In general, cultures derived from embryonic tissues survive and proliferate better than those from the adult. This presumably reflects the lower level of specialization and higher proliferative potential in the embryo. Adult tissues usually have a lower growth fraction and a higher proportion of non-replicating specialized cells, often within a more structured and less readily disaggregated extracellular matrix. Initiation and propagation are more difficult, and the lifespan of the culture often shorter.

Embryonic or fetal tissue has many practical advantages, but it must be remembered that in some instances the cells will be different from adult cells and it cannot be assumed that they will mature into adult-type cells unless this can be confirmed by appropriate characterization.

Examples of widely used embryonic cell lines are the various 3T3 lines (primitive mouse mesodermal cells) and WI-38, MRC-5, and other human fetal lung fibroblasts. Mesodermally-derived cells (fibroblasts, endothelium, myoblasts) are on the whole easier to culture than epithelium, neurons, or endocrine tissue. This selectivity may reflect the extensive use of fibroblast cultures during the early development of culture media and the response of mesodermallyderived cells to mitogenic factors present in serum. Selective media have now been designed for epithelial and other cell types, and with some of these it has been shown that serum is inhibitory to growth and may promote differentiation. Primary culture of epithelial tissues, such as skin, lung, and mammary gland, is routine in some laboratories, and prepared cultures are available commercially.

Normal or neoplastic?

Normal tissue usually gives rise to cultures with a finite lifespan, while cultures from tumours can give rise to continuous cell lines. However, there are several examples of continuous cell lines (BHK-21 hamster kidney fibroblasts, MDCK dog kidney epithelium, 3T3 fibroblasts) which have been derived from normal tissues and which are non-tumorigenic.

Normal cells generally grow as an undifferentiated stem cell or precursor cell, and the onset of differentiation is accompanied by a cessation in cell proliferation which may be permanent. Some normal cells, such as fibrocytes or endothelium, are able to differentiate and still de-differentiate and resume proliferation and in turn re-differentiate, while others, such as squamous epithelium, skeletal muscle, and neurons, once committed to differentiate, appear to be incapable of resuming proliferation.

Cells cultured from neoplasms, such as B16 mouse melanoma, can express at least partial differentiation, while retaining the capacity to divide. Many studies of differentiation have taken advantage of this fact and used differentiated tumours such as hepatomas and human and rodent neuroblastomas, although whether this differentiation is normal is not known.

Tumour cells can often be propagated in the syngeneic host, providing a cheap and simple method of producing large numbers of cells, albeit with lower purity. Where the natural host is not available, tumours can also be propagated in animals with a compromised immune system, with greater difficulty and cost, but similar advantages. Many other differences between normal and neoplastic cells are similar to those between finite and continuous cell lines as immortalization is an important component of the process of transformation.

Subculture

Freshly isolated cultures are known as *primary cultures* until they are subcultured. They are usually heterogeneous and have a low growth fraction, but are more representative of the cell types in the tissue from which they were derived and in the expression of tissue-specific properties. Subculture allows the expansion of the culture (it is now known as a cell line), the possibility of cloning, characterization and preservation, and greater uniformity, but may cause a loss of specialized cells and differentiated properties unless care is taken to select the correct lineage and preserve or reinduce differentiated properties. The greatest advantage of subculturing a primary culture into a cell line is the provision of large amounts of consistent material suitable for prolonged use.

Finite or continuous cell lines?

After several subcultures a cell line will either die out (finite cell line) or 'transform' to become a continuous cell line. It is not clear in all cases whether the stem line of a continuous culture pre-exists or arises during serial propagation. Because of the time taken for such cell lines to appear (often several months) and the differences in their properties, it has been assumed that a mutation occurs, but the pre-existence of immortalized cells, particularly in cultures from neoplasms, cannot be excluded. Complementation analysis has shown that senescence is a dominant trait, and immortalization the result of mutations and/or deletions in genes such as p53. The activity of telomerase is higher in immortal cell lines and may be sufficient to endow the cell line with an infinite lifespan.

The appearance of a continuous cell line is usually marked by an alteration in cytomorphology (giving cells that are smaller, less adherent, more rounded, and with a higher ratio of nucleus to cytoplasm), an increase in growth rate (population doubling time can decrease from 36-48 h to 12-36 h), a reduction in serum dependence, an increase in cloning efficiency, a reduction in anchorage dependence (as measured by an increased ability to proliferate in suspension as a liquid culture or cloned in agar), an increase in heteroploidy (chromosomal variation among cells) and aneuploidy (divergence from the donor, euploid, karyotype), and an increase in tumorigenicity. The resemblance between spontaneous *in vitro* transformation and malignant transformation is obvious but nevertheless the two are not necessarily identical although they have much in common. Normal cells can 'transform' to become continuous cell lines without becoming malignant, and malignant tumours can give rise to cultures which '*transform*' and become more (or even less) tumorigenic, but acquire the other properties listed above. Transformation *in vitro* is primarily the acquisition of an infinite lifespan. Simultaneous or subsequent alterations in growth control can be under positive control by oncogenes or negative control by tumour suppressor genes.

The advantages of continuous cell lines are their faster growth rates to higher cell densities and resultant greater yield, their lower serum requirement and general ease of maintenance in commercially available media, and their ability to grow in suspension. Their disadvantages include greater chromosomal instability, divergence from the donor phenotype, and loss of tissue-specific markers.

A number of techniques, including transfection or infection with viral genes (such as E6 and E7 from human papilloma virus, and SV40T) or viruses (such as *Epstein-Barr virus*, EBV), have been used to immortalize a wide range of cell types. The retention of lineage-specific properties is variable.

Propagation in suspension

Most cultures are propagated as a monolayer attached to the substrate, but some, including transformed cells, haematopoietic cells, and cells from ascites, can be propagated in suspension. Suspension culture has advantages, including simpler propagation (subculture only requires dilution, no trypsinization), no requirement for increasing surface area with increasing bulk, ease of harvesting, and the possibility of achieving a '*steady state*' or biostat culture if required.

Selection of Medium

Regrettably, the choice of medium is still often empirical. What was used previously by others for the same cells, or what is currently being used in the laboratory for different cells, often dictates the choice of medium and serum. For continuous cell lines it may not matter as long as the conditions are consistent, but for specialized cell types, primary cultures, and growth in the absence of serum, the choice is more critical.

There are two major advantages of using more sophisticated media in the absence of serum: they may be selective for particular types of cell, and the isolation of purified products is easier in the absence of serum. Nevertheless, culture in the presence of serum is still easier and often no more expensive, though less controlled. Two major determinants regulate the use of serum-free media:

1. Cost. Most people do not have the time, facilities, or inclination to make up their own media, and serum-free formulations, with their various additives, tend to be much more expensive than conventional media.
2. Requirements for serum-free media are more cell type specific. Serum will cover many inadequacies revealed in its absence. Furthermore, because of their selectivity, a different medium may be required for each type of cell line. This problem may be particularly acute when culturing tumour cells where cell line variability may require modifications for cell lines from individual tumours.

In the final analysis the choice is often still empirical: read the literature and determine which medium has been used previously. If several media have been used, as is often the case, test them all, with others added if desired. Measure the growth (*population doubling time* (PDT) and *saturation density*), cloning efficiency, and expression of specific properties (differentiation, transfection efficiency, cell products, etc.). The choice of medium may not be the same in each case, for example differentiation of lung epithelium will proceed in serum, but propagation is better without. If possible, include one or more serum-free media in the panel to be tested, supplemented with growth factors, hormones, and trace elements as required.

Once a medium has been selected, try to keep this constant for as long as possible. Similarly, if serum is used, select a batch by testing samples from commercial suppliers and reserve enough to last six months to one year, before replacing it with another pre-tested batch. Testing procedures are as described above for media selection.

Gas Phase

The composition of the gas phase is determined by:

1. The type of medium (principally its sodium bicarbonate concentration);
2. Whether the culture vessel is open (Petri dishes, multiwell plates) or sealed (flasks, bottles); and
3. The amount of buffering required.

Several variables are in play, but one major rule predominates and three basic conditions can be described. The rule is that the bicarbonate concentration and carbon dioxide tension must be in equilibrium. It should be remembered that carbon dioxide/ bicarbonate is essential to most cells, so a flask or dish cannot be vented without providing carbon dioxide in the atmosphere. Prepare medium to about pH 7.1-7.2 at room temperature, incubate a sample with the correct carbon dioxide tension for at least 0.5 h in a shallow dish, and check that the pH stabilizes at pH 7.4. Adjust with sterile 1 M HC1 or 1 M NaOH if necessary.

Oxygen tension is usually maintained at atmospheric pressure, but variations have been described, such as elevated for organ cultureand reduced for cloning melanoma aad some haematopoietic cells.

Culture System

Originally, tissue culture was regarded as the culture of whole fragments of explanted tissue with the assumption that histological integrity was at least partially maintained. Now '*tissue culture*' has become a generic term and includes organ culture, where a small fragment of tissue or whole embryonic organ is explanted to retain tissue architecture, and cell culture, where the tissue is dispersed mechanically or enzymatically, or the cells migrate from an explant, and the cells are propagated as a suspension or attached monolayer.

Cell cultures are usually devoid of structural organization, have lost their histotypic architecture and often the biochemical properties associated with it, and generally do not achieve a steady state unless special conditions are employed. They can, however, be propagated, expanded, and divided into identical replicates. They can be characterized and a denned cell population preserved by freezing, and they can be purified by growth in selective media, physical cell separation, or cloning, to give a characterized cell strain with considerable uniformity.

Organ culture will preserve cell interaction and retain histological and biochemical differentiation for longer than cell culture. After the initial trauma of explantation and some central necrosis, organ cultures can remain in a steady state for a period of several days to years. However, they cannot be propagated easily, show greater experimental variation between replicates, and tend to be more difficult to use for quantitative determinations.

Purified cell lines can be maintained at high cell density to create histotypic cultures and different cell populations can be combined in organotypic culture, simulating some of the properties of organ culture.

Substrate

The nature of the substrate is determined largely by the type of cell and the use to which it will be put. Polystyrene which has been treated to make it wettable and give it a net negative charge is now used almost universally. In special cases the plastic is pre-coated with fibronectin, collagen, gelatin, or poly-L-lysine. Glass may also be used, but must be washed carefully with a non-toxic detergent.

Scale

Culture vessels vary in size from microtitration (30 mm^2, 100-200 μl) up through a range of dishes and flasks to 180 cm^2, and roller bottles and multisurface propagators for large scale culture. The major determinants are the number of cells required (5×10^4 to 10^5/cm^2 maximum for most untransformed cells, 10^5 to 10^6/cm^2 for transformed), the number of replicates, and the times of sampling: a 24-well plate is good for a large number of replicates for simultaneous sampling, but individual dishes, tubes, or bottles are preferable where sampling is carried out at different times. Petri dishes are cheaper than flasks and good for subsequent processing, e.g. staining or extractions. Flasks can be sealed, do not need a humid carbon dioxide incubator, and give better protection against contamination. Volume is the main determinant for suspension cultures. Sparging and agitation will become necessary as the depth increases. Where a product is required rather than cells, there are advantages in perfusion systems which can be used to culture cells on membranes or hollow fibres. These supply nutrients across the membrane, and the product is collected either from the cell supernatant or the medium, depending on the molecular weight, which will determine the partitioning of the product on either side of the membrane. Perfusion is also useful for time lapse studies, where cells are monitored on a microscope stage by a video camera, and for pharmacokinetic modelling where the duration and concentration of a test compound can be regulated precisely.

PROCEDURES

Substrate

Most laboratories now utilize disposable plastics as substrates for tissue culture. They are optically clear, prepared for tissue culture use by modification of the plastic to make it wettable and suitable for cell attachment, and come sterilized for use. On the whole they are convenient and provide a reproducible source of vessels for both routine and experimental work.

Some more fastidious cell types, such as bronchial epithelium, vascular endothelium, skeletal muscle, and neurons require the substrate to be coated with extracellular matrix materials such as fibronectin, collagen or laminin. Most matrix products are available individually or combined in Matrigel, extracellular matrix produced by the Engelberth Holm Swarm sarcoma cell line. Alternatively, the substrate may be coated with extracellular matrix by growing cells on the plastic and then washing them off with 1% Triton X in ultrapure water.

Matrix coating can be carried out in three ways:

1. By wetting the surface of the plastic with the matrix component(s), incubating for a short period (usually ~ 30 min), then removing the surplus and using the plastic with adsorbed matrix within seven to ten days (stored at 4°C if not used immediately).
2. By wetting the plastic and removing the surplus matrix material and allowing the residue to dry.
3. By adding collagen or Matrigel and allowing it to gel.

Wet or dry coating is used mainly for propagation, while gel coating is used to promote differentiation of cells growing on or in the gel.

Medium

Most of the commonly used media are available commercially, presterilized. For special formulations or additions it may be necessary to prepare and sterilize some of the constituents. In general, stable solutions (water, salts, and media supplements such as tryptose or peptone) may be autoclaved at 12°C (100 kPa or 1 atmosphere above ambient) for 20 min, while labile solutions (media, trypsin and serum) must be filtered through a 0.2 μm porosity membrane filter. Sterility testing should be carried out on samples of each filtrate.

Where an automatic autoclave is used care must be taken to ensure that the timing of the run is determined by the temperature of the centre of the load and not just by drain or chamber temperature or pressure which will rise much faster than the load. The recorder probe should be placed in a package or bottle of fluid similar to the load and centrally located.

Cell Culture

Primary cultures

The first step in preparing a primary culture is sterile dissection followed by mechanical or enzymatic disaggregation. The tissue may simply be chopped to around 1 mm^3 and the pieces attached to a dish by their own adhesiveness, by scratching the dish, or by using clotted plasma. In these cases cells will grow out from the fragment and may be used directly or subcultured. The fragment of tissue, or explant as it is called, may be transferred to a fresh dish or the outgrowth trypsinized to leave the explant and a new outgrowth generated.

When the cells from the outgrowth are trypsinized and reseeded into a fresh vessel they become a secondary culture, and the culture is now technically a cell line.

Primary cultures can also be generated by disaggregating tissue in enzymes such as trypsin (0.25% crude or 0.01-0.05% pure) or collagenase (200-2000 U/ml, crude) and the cell suspension allowed to settle on to, adhere, and spread out on the substrate. This type of culture gives a higher yield of cells though it can be more selective, as only certain cells will survive dissociation. In practice, many successful primary cultures are generated using enzymes such as collagenase to reduce the tissue, particularly epithelium, to small clusters of cells which are then allowed to attach and grow out.

When primary cultures are initiated, all details of procedures should be carefully documented to form part of the provenance of any cell line that may arise and be found to be important. A sample of tissue, or DNA extracted from it, should be archived to be available for DNA fingerprinting or profiling for authentication of any cell lines that arise.

Subculture

A monolayer culture may be transferred to a second vessel and diluted by dissociating the cells of the monolayer in trypsin (suspension cultures need only be diluted). This is best done by rinsing the monolayer with PBS lacking Ca^{2+} and Mg^{2+} (PBSA) or PBSA containing 1 mM EDTA, and removing the rinse, adding cold trypsin (0.25% crude or 0.01-0.05% pure) for 30 sec, removing the trypsin, and incubating in the residue for 5-15 min, depending on the cell line. Cells are then resuspended in medium, counted and reseeded.

Growth curve

When cells are seeded into a flask they enter a lag period of 2-24 h, followed by a period of exponential growth (the '*log phase*'), and finally enter a period of reduced or zero growth after they become confluent ('*plateau phase*'). These phases are characteristic for each cell line and give rise to measurements which should be reproducible with each serial passage: the length of the lag period, the *population*

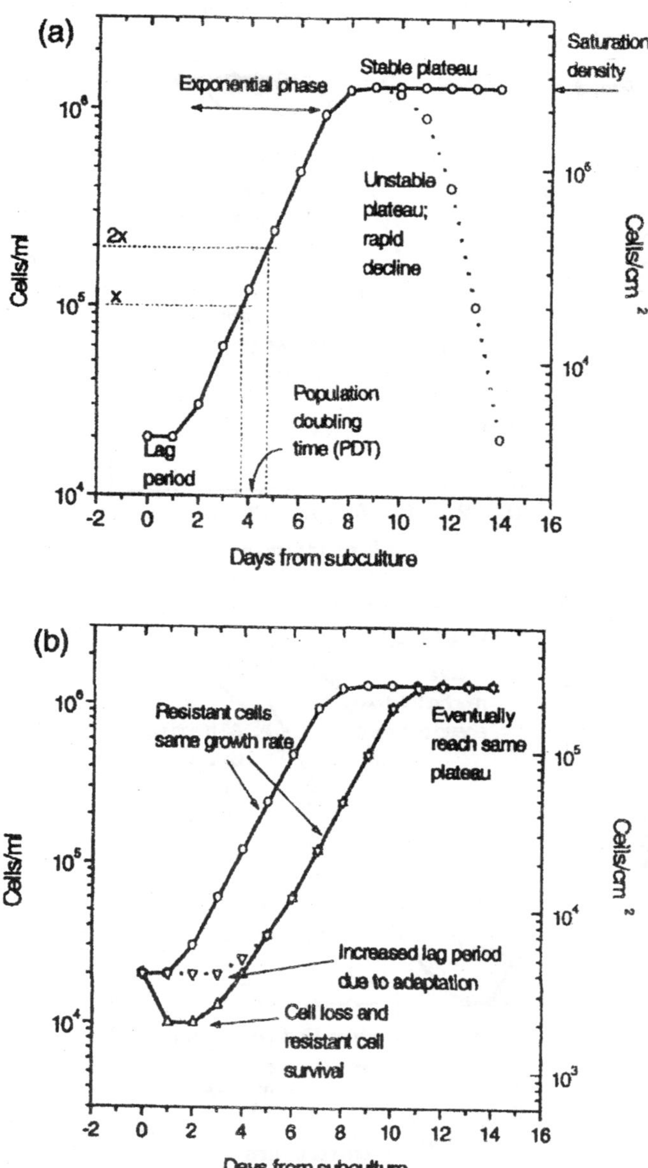

Fig. 1.1. Analysis of growth curves.

doubling time (PDT) in mid-log phase, and the saturation density at plateau, given that the environmental conditions are kept constant. They should be determined when first handling a new cell line and at

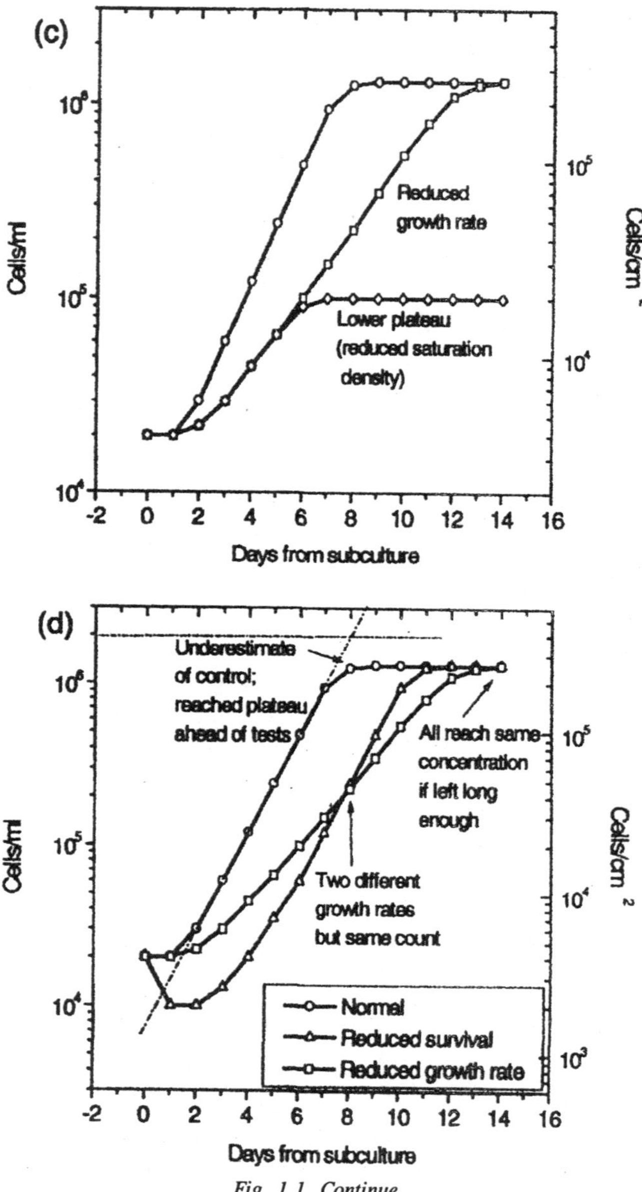

Fig. 1.1. Continue

intervals of every few months thereafter. It is an important element of quality control to be able to demonstrate that the same seeding concentration will yield a reproducible number of cells at subculture, carried out after a consistent time interval, without necessarily performing a growth curve each time.

The determination of the growth cycle is important in designing routine subculture and experimental protocols. Cell behaviour and biochemistry changes significantly at each phase and it is therefore essential to control the stage of the growth cycle when drugs or reagents are added or cells harvested. The shape of the growth curve can also give information on the reproductive potential of the culture where differences in growth rate (PDT), adaptation or survival, and density limitation of growth (level of saturation density in plateau), can be deduced from the shape of the curve. However it is generally recognized that the analysis of clonal growth is easier, and less prone to ambiguity and misinterpretation.

Feeding

Some rapidly growing cultures, such as transformed cell lines like HeLa, will require a medium change after three to four days in a seven day subculture cycle. This is usually indicated by an increase in acidity where the pH falls below pH 7.0. Medium can also deteriorate without a major pH change, as some constituents, like glutamine, are unstable, and others may be utilized without a major pH shift. It is, therefore, recommended that the medium is changed at least once per week.

Contamination

The problem of microbial contamination has been greatly reduced by the use of laminar airflow cabinets. The risk of contamination can also be reduced by use of antibiotics, but this should be reserved for high-risk procedures, such as primary culture, and cultures should be maintained in the absence of antibiotics so that chronic, cryptic contaminations are not harboured. Check frequently for contamination by looking for a rapid change in pH (usually a fall, but some fungi can increase the pH), cloudiness in the medium, extracellular granularity under the microscope, or any unidentified material floating in the medium. If a contamination is detected, discard the flask unopened and autoclave. If in doubt, remove a sample and examine by phase microscopy, Gram's stain, or standard microbiological techniques.

Mycoplasma

Cultures can become contaminated by mycoplasma from media, sera, trypsin, imported cell lines, or the operator. Mycoplasma are not visible to the naked eye, and, while they can affect cell growth, their presence is often not obvious. It is important to test for mycoplasma at regular intervals (every one to three months) as they can seriously affect almost every aspect of cell metabolism, antigenicity, and growth characteristics. Several tests have been proposed, but the fluorescent DNA stain technique of Chen is the most widely used, although it is less sensitive than the PCR-based and culture methods.

Cross-contamination

Its severity is often underrated, and consequently it still occurs with a higher frequency than many people admit. A significant number of cell lines, including Hep-2, KB, and Chang liver, are still in regular use without acknowledgement of their contamination with HeLa. Many other cell lines are cross-contaminated with cells other than HeLa. Many more cross-contaminations are yet to be detected.

To avoid cross-contamination:

1. Do not share bottles of media or reagents among cell lines;
2. Do not return a pipette, which has been in or near a flask or bottle containing cells, back to the medium bottle; use a fresh pipette;
3. Do not share medium among operatives; and
4. Handle one cell line at a time.

Instability and preservation

Early passage cell lines are unstable as they go through a period of adaptation to culture. However, between about the 5th and 35th generation (for human diploid fibroblasts; other cell types may be different) the culture is fairly consistent. As the culture will start to senesce as it gets older, finite cell lines should be used after the period of adaptation but before senescence. As continuous cell lines can be genetically unstable they should only be used continuously for approximately three months, before stock replacement. Some cell lines, such as 3T3 and other mouse lines which are immortal but not transformed, can transform spontaneously, and should not be propagated for prolonged periods.

Validated and authenticated frozen stocks of all cell lines should be maintained to protect against cell line instability, and to give insurance against contamination, incubator failure, or other accidental loss. Animal cell culture is a precise discipline. Beware of those who say it is not, or it is '*magic*', or due to '*green fingers*'; they are not controlling all the variables. Consistency can be achieved, and the following chapters are intended to indicate how best to control conditions within the present limitations of our knowledge.

2

LABORATORY METHODS

In this chapter the subject of scale-up is reviewed, which is taking small laboratory cultures (e.g. 10 ml) to industrial-scale processes (e.g. 10,000 litre), i.e. a 1,000,000-fold scale-up! The aim of such scale-up is to provide more cells, and more cell product, in as efficient and cost-effective a manner as possible. Cell cultures have been used since 1954 for the production of human (e.g. polio, measles, mumps, rabies, rubella) and then veterinary (e.g. FMDV) vaccines. Interferon was the next most important product to be developed, followed by monoclonal antibodies and a range of recombinant proteins.

HISTORY

The importance of cell culture as a manufacturing process is exemplified by the range of products either licensed or in trial. This production activity has been made possible by the ability to grow cells in unit processes of 10^{11}-10^{13} cells. The scale-up process involves many factors, which include engineering (especially to maintain sterility in large-scale plants), physiological (supply of oxygen and nutrients and maintenance of correct environmental conditions), monitoring and control (biosensors, performance measurements, computer control) and logistic and regulatory requirements (compliance with pharmaceutical licensing and health and safety aspects when growing pathogens).

Scale-up Factors

The primary aim has been to move from multiple processes (many replicate bottles, roller cultures, spinner flasks, etc.) to a high-volume unit process. This has economical advantages in terms of manpower needed (reduction in repetitive steps for each culture) and space (particularly expensive hot-room facilities), and efficiency advantages in that it is feasible to control, reproducibly, environmental factors such as pH and O_2 in a unit process. One cannot help being aware of the huge range of alternative culture systems that are available. These are a result of solutions to overcoming limitations to scale-up and the need to find solutions for both suspension and anchorage-dependent cells.

The first problem on moving from a small multiple-process unit to a large unit process is usually oxygen limitation. To overcome this and supply oxygen without the damaging effects of air bubbles from sparging and the high stirring speeds needed to effect efficient mass transfer, bioreactors have been developed that oxygenate through a membrane inside the bioreactor or by external medium loops through hollow-fibre oxygenators. A related factor is mixing, and low-shear modifications to impellers or the airlift reactor concept have been introduced. A problem for anchorage-dependent cells is the limitation on surface area during scale-up. A huge range of reactors have been developed (multiple plates, spirals, ceramic matrices, glass beads) but the most successful has been the microcarrier. By growing cells on a small bead, which can be cultured analogously to a suspension cell, the large fermenter systems used for suspension cells have been made available for anchorage-dependent cells. The next problem to be overcome is nutrient limitation and toxic metabolite build-up. Although medium changes can be used (especially for attached cells), or detoxification procedures, the most efficient means of overcoming these problems is continuous medium perfusion. To perfuse suspension cells, physical separation of cells and waste medium is necessary: this has been done by means of spin-filters; external loops to filters or centrifuges; or gravitational settling devices. The use of perfusion resulted in the development of many immobilization procedures to prevent cell wash-out (e.g. encapsulation, hollow-fibre devices, membrane reactors, gels, fibres and porous microcarriers). Additional benefits to these developments were the provision of huge unit surface areas for cell attachment and the ability to grow cells at 50-100-fold higher densities than previously achieved in free suspension.

Scale-up Strategies

1. High-volume (to 10,000 l) tank fermenter systems with conventional cell densities of 1-2 $\times$ 10^6 ml^{-1}.
2. Low-volume (under 1 l) but very high density (1-2 $\times$ 10^8 ml^{-1}) heterogeneous systems (e.g. hollow-fibre bioreactors).
3. Intermediate systems using a spin-filter to increase cell density to 1-2 $\times$ 10^7 ml^{-1} but to a volumetric scale of 500 l only.

4. High-density (1×10^8 ml^{-1}) scaleable systems (to 200 l) based on porous micro-carriers.

The high-density immobilized systems have many advantages in terms of smaller process equipment, higher product concentration and reduced nutritional requirements, and perfusion allows long culture runs (thus reducing the down-time of the plant) and shorter residence times for products. They also have the advantage that they are equally suitable for suspension and anchorage-dependent cells, they provide protection from shear and they have a three-dimensional structure that enhances productivity. Basically, the low-volume, high-density systems fill an important niche in manufacturing capability (e.g. gramme amounts of monoclonal antibodies), whilst the high-volume systems are the most widely used scale-up route for pharmaceutical products. However, there is increasing acceptance of, and need for, the scaleable high-density systems and these will be the first-choice system in future years.

General Principles

At some stage a decision has to be made, if scale-up is required, of when to move from laboratory units to small-scale industrial systems. This is a decisive moment because of the investment needed to set up *in situ* fermentation systems. This step should be taken at 10 l and involves:

1. Change from glass to stainless-steel vessels.
2. Change from a mobile to a static (plumbed-in) system with connection to steam for *in situ* sterilization, manipulation being carried out in the open laboratory (not a laminar flow cabinet), more sophisticated temperature control and additional vessels for medium holding and culture harvesting.
3. More sophisticated and sensitive environmental control systems.

A typical cell culture bioreactor available from all fermenter suppliers is conceptually similar to the familiar bacterial fermenter apart from modifications such as a marine impeller, curved or convex base for better mixing at low speeds and water-jacket—not immersion heater - temperature control.

Monolayer and Suspension Culture

Suspension culture is the preferred method for scale-up because it is easier and cheaper, it requires less space, cell growth can be monitored and environmental parameters can be controlled more easily. However, many cell lines have higher specific productivity when attached to the substrate. In the urge to move to suspension systems, the following advantages of anchorage-dependent systems should not be overlooked:

1. They facilitate complete medium changes, which are often needed, for example, to wash out serum before adding serum-free production medium.
2. They are easier to perfuse
3. Medium/cell ratios are changed more easily during an experiment

Microcarrier culture has most of the advantages of both suspension and anchorage-dependent systems and this is a particularly valuable asset for porous microcarrier culture.

Culture Modes

1. *Batch culture:* cells grow from seed to a final density over 4-7 days and are then harvested (typical monoclonal antibody yield is 6 mg l^{-1} per batch run).
2. *Fed-batch culture:* additional media, or medium components, are added to increase the culture volume and density of cells.
3. *Semi-continuous batch culture:* this is in effect a batch culture that is partially harvested (e.g. 70%), topped up with fresh medium, allowed to grow up and then harvested again. Typically, three or four harvests can be made, although usually with diminishing returns at each harvest.
4. *Continuous-flow (chemostat) culture:* the culture is completely homogeneous for long periods of time whilst cells are in a steady state. An extremely useful tool for physiological studies but not very economical for production (by definition, cells are never at maximum density)
5. *Continuous perfusion culture:* differs from chemostat in that all nutrients are kept in excess and cells are retained within the hioreactor. The most productive culture system gives typical yields of 10 mg l^{-1} day^{-1} for 50-100 days.

There is no definitive answer as to the best system to use: it depends upon the nature of the cell and the product, the quantity of product, downstream processing capability, licensing regulations, etc. However, a rough guide to relative costs for producing monoclonal antibodies by perfusion, continuous-flow and batch culture is the ratio 1:2:3:5.

Biological Factors

Key factors for the successful scale-up of cell culture include using the best available cell line, inoculating cells that are in prime

Table 2.1. Comparison of culture systems indicating volumetric size, unit and system cell yields

Scale (l)	*System*	*Area (cm²) (batch size)*	*Cells per batch (cells l^{-1})*
	Roller	1750 (× 100)	2×10^{10} (2×10^{8})
	Multitray	24,000 (× 10)	3×10^{10} (3×10^{9})
	CellCube	85,000 (× 4)	4×10^{10} (9×10^{9})
1	Hollow fibre	20,000	10^{10} (10^{11})
20	Microporous microcarrier	2.5×10^{6}	10^{12} (5×10^{10})
100	Glass sphere	10^{6}	10^{11} (10^{9})
200	Stacked plate	2×10^{5}	2×10^{10} (1×10^{8})
500	Microcarrier (with spin-filter)	2×10^{10}	10^{13} (5×10^{10})
2000	Airlift		4×10^{12} (2×10^{9})
4000	Microcarrier	5×10^{6}	8×10^{12} (2×10^{9})
10,000	Stirred tank		2×10^{13} (2×10^{9})

condition and ensuring that nutritional, physiological and physico-chemical conditions are optimized.

Cells vary in their cultural characteristics. This is very noticeable when comparing clones from the same hybridization or transfection experiment. Differences in expression of the desired product are well recognized, hence the standard screening produced for high producers. However, there is also clonal variability in serum requirements, plating efficiency, growth rate, metabolism (e.g. glycolytic rate) and ability to grow in (or adapt to) suspension culture. It is thus beneficial to select a clone that has the desired characteristics for the scaled-up process that is to be used.

The quicker that cells can be harvested and re-inoculated into a new culture, the healthier they tend to be and the higher the chance of establishing a successful culture. Prolonged exposure to proteolytic enzymes (e.g. trypsin), standing in concentrated suspensions for long periods before inoculation and over-robust mixing (or pouring) whilst in the fragile post-trypsinized state must be avoided. Selecting cells in the late logarithmic phase and rapid processing into the new culture at an adequate inoculum level should allow an optimum initiation of the new culture with a short lag phase and maximum cell density. Always inoculate cells into stabilized culture conditions, because shifts in temperature, pH and oxygen levels during initiation can be damaging.

As one scales up, culture conditions become more demanding on the cells and the logistics of preparing large-cell inocula and maintaining sterility become more difficult. Therefore, do not be over-ambitious in the scale-up steps.

In this overview an explanation has been given for the wide and complex range of cell culture bioreactor systems that are available. This is to put some perspective on the ones that are described in more detail in the following chapters. The selection of a suitable system follows the adage '*horses for courses*', i.e. a process is selected that fulfils the particular criteria for the cell, product, scale and quantity required, plus the resources in facilities, manpower and experience that are available.

There are some additional general factors to he taken into consideration. First, upstream processing is only part of the total process, which starts at the initiation of a cell line and finishes with a purified and packaged product, and therefore should not be treated in isolation. Secondly, increasing unit productivity does not only depend upon designing a bigger, better and more efficient bioreactor. Significant increases in productivity have been, and will be, achieved by designing better media, genetically constructing more efficient and robust cell lines and more critical monitoring (by biosensors) and control (by computer) of the process.

Roller Bottle Culture

Roller bottle culture is considered the first scale-up step for anchorage-dependent cells from stationary flasks or bottles. This is achieved by using all the internal surface for cell growth, rather than just the bottom of a bottle. The added advantages are that: a smaller volume of medium and thus a higher product titre can be achieved; the cells are more efficiently oxygenated due to alternative exposure to medium and the gas phase; and dynamic systems usually generate higher unit cell densities than stationary systems.

The basis of the roller culture system is to place multiple cylindrical bottles into an apparatus that will rotate the bottles evenly at set rotational speeds (between 5 and 60 rph). Apparatus is available that will accommodate four to hundreds of bottles, and in fact this method is used for the large-scale production of vaccines using 29,000 bottles (1-l volume) per batch. Any cylindrical bottle of the correct tissue culture grade surface (borosilicale glass or polystyrene) can be used and there are many available commercially between 250 and 2000 cm^2 in capacity. It is an advantage when using large numbers to have as small a diameter as possible in order to reduce the total volume of the culture vessels. However, sufficient headspace must be maintained to prevent oxygen limitation, and thus the medium/air volume is usually between 1:5 and 1:10.

Procedure: Roller Bottle Culture of Animal Cells

Materials and equipment

- Standard tissue culture media and reagents
- Roller bottles (e.g. 23 × 12-cm plastic bottle with surface area of 1400 cm^2 from, for example, Becton Dickinson, Sterilin).
- Roller bottle apparatus with speed control between 5 and 60 rph.

1. Add approximately 1 ml of complete culture medium at 37°C per 5-cm^2 culture area.
2. Add 5% CO_2 in air to headspace.
3. Add 1-2 × 10^5 cells cm^{-2}.
4. Seal bottle and place in roller culture apparatus.
5. Rotate bottle at 12-24 rph for the initial attachment phase (2-8 h, depending upon cell type). This faster speed is to get an even distribution of cells but should be reduced for cells with low attachment efficiency.
6. Reduce revolution rate to 5-10 per hour when culture is growing.
7. Cell growth can be monitored initially under an inverted microscope and later in the culture period by visual inspection.
8. A medium change can be carried out after 4-5 days if the pH becomes acid and/or maximum cell densities are required. This can also be carried out to change to a production medium, or when infecting cells, and a lower medium volume can be added to get a higher product concentration.
9. Harvest cells when confluent (5-6 days) by removing the medium and trypsinizing in the conventional way. After adding trypsin, roll the bottles at speeds of 20-60 rph until cells detach. Cell yields will be similar to or up to two-fold higher than in stationary cultures, and multilayering (non-diploid cell lines) will occur.

Comment

Successful roller bottle culture depends a great deal on solving various logistic problems. Larger roller bottles (1500 cm^2) are unwieldy to use and cannot he manipulated in many tissue culture cabinets. Special Class II cabinets are suitable; otherwise use the smaller size bottles. Also, do not underestimate the time involved in harvesting cells from roller bottles. If a large number of bottles are being used, trypsinize four at a time and get the cells in fresh medium with serum before beginning the next batch. Cells deteriorate very rapidly in trypsin, and also when being held at high concentrations in medium, so do not store whilst harvesting multiple batches of roller bottles, but put them in a new culture as soon as possible. Unless your laboratory is geared up for this type of work (multiple tissue culture cabinets and staff), limit the number of roller bottles to 8-16 at a time. It is very difficult to remove completely all the cells from a roller bottle without repeated washing (increases contamination risk, standing time of cells and centrifugation volume) and it is wise to assume that a 10-20% loss may occur. This is particularly important when judging how many bottles to set up as an inoculum for a larger culture.

Supplementary Procedures

The roller bottle system is a multiple process requiring considerable staff time for repeated manipulations. Thus many modifications have been introduced that increase the efficiency or surface area capacity of roller bottles. Some examples are follows:

1. *Perfusion*. Specially modified swivel caps have been developed that allow continuous perfusion of roller bottles.
2. *Spira-cell multi-surface roller bottle*. This bottle contains a spiral polystyrene cartridge on which cells grow on both sides of the film; 3000-cm^2, 4500-cm^2 and 6000-cm^2 sizes are available.
3. *Extended surface area roller bottle* (ESRB). This polystyrene bottle has a ribbed or corrugated surface, doubling the surface area of the 850-cm^2 flask to 1700 cm^2.
4. *Bellco-corbeil culture system*. A roller bottle is packed with a cluster of small glass tubes (arranged in parallel and separated by silicone spacer rings). Models are available in 1000-cm^2, 10,000-cm^2 and 15,000-cm^2 sizes. Medium is perfused through the vessel and the bottles rotate 360° in alternate directions to avoid tube twisting and the need for special caps.
5. If hundreds of roller bottles need to he handled, then a robotic culture system could be considered, which reduces manning by 10-fold as well as giving improved consistency and reproducibility.

The roller bottle technique is a well-established and successful culture method widely used for the production of cells and products. One reason for this is that a single contamination event does not mean that the whole batch is lost, as with a single unit process. However, roller culture needs a considerable financial investment both in the apparatus itself and in incubator/hot-room facilities. Also, some cell lines (particularly epithelial) may not be as successfully grown in roller bottles as in stationary bottles. Common problems are streaking, clumping or inadequate spreading over the total surface (e.g. non-locomotory cell lines). An alternative scale-up route is to use multi-surface stationary systems such as the Nunclon Cell Factory (Nunc) or the CellCube (Costar).

Spinner Flask Culture

The first scale-up step for cells growing in suspension either naturally or after adaptation, or anchorage-dependent cells on microcarriers, is the spinner flask. This technique originated by placing a silicone or Teflon-coated magnet with a central ring into a glass vessel that is placed on a magnetic stirrer. Although this simplistic approach is still used, it is preferable to use a specially constructed flask where the magnet is suspended (essential for microcarrier culture, because otherwise the cells and microcarriers will be destroyed by the grinding action of the magnet) just above the bottom of the flask. In addition, side arms are fitted in order to add cells, change medium or gas with oxygen or CO_2-enriched air (in which cases filters should be fitted). It is preferable to have a slightly convex surface under the stirrer bar to aid mixing. There are many variations on this theme from different suppliers, mostly with regard to size, number and positioning of the stirrer bar.

The spinner flask is a glass vessel, usually intended to be used for replicate cultures, on a multibased magnetic stirrer. Sizes range from a few millilitres to 20 l, but for ease of handling, safety and physicochemical reasons it is advisable to consider 10 l as a maximum practical size. The units are sterilized by autoclaving. A very important factor is the quality of the magnetic stirrer. This should be able to give an extremely stable speed of rotation between 10 and 300 rpm, have an accurate tachometer, restart after power failure at the set speed, not overheat on the top surface and operate reliably for long periods of time in a 37°C environment. Thus high-quality stirrers are essential with a strong magnetic field. For unit scales over 10 l, and definitely for those over 20 l, *in situ* fermentation systems should be used rather than spinner flasks.

Procedure: Culture of Suspension Cells in a Spinner Flask

Reagents and solutions

Growth medium

Standard tissue culture media can be used with various supplements. Serum at 10% can cause foaming at fast stirring speeds, so the serum

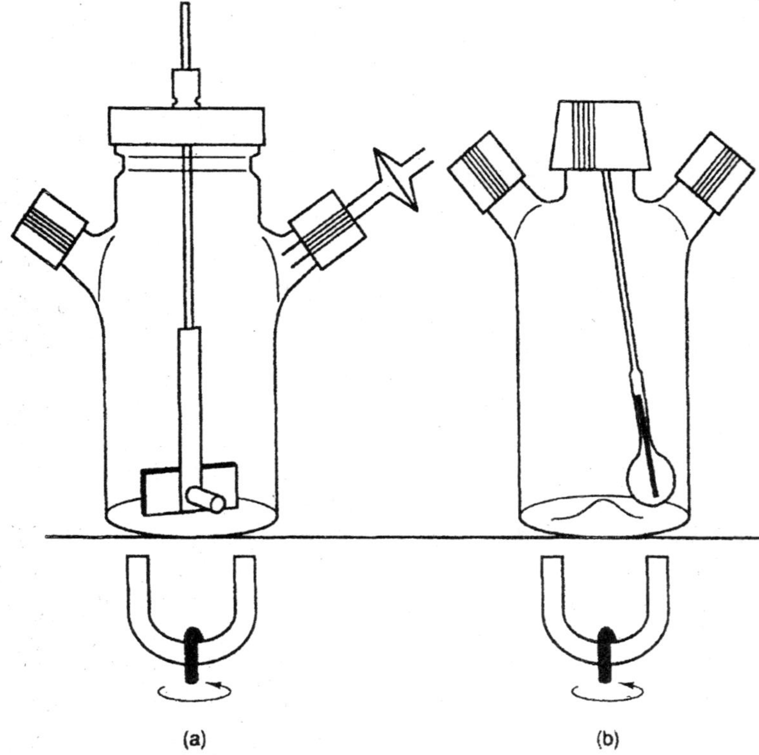

Fig. 2.1. (a) A typical spinner flask with magnetic spinner bar. (b) A Techne flask with asymmetric stirring rod.

level is usually reduced to as low a level as possible (0.5-5%). The use of HEPES buffer to stabilize the pH during the setting-up procedure is beneficial.

Medium supplements

Pluronic F-68 (polyglycol) (BASF, Wyandot) can be used at 0.1% to protect cells against mechanical damage, especially at reduced serum concentrations. *Carboxymethyl cellulose* (CMC) (15-20 cP) can also be used to protect cells from mechanical damage at 0.1% concentration. Antifoam (6 ppm) is recommended at serum concentrations above 2-3%.

Materials and equipment

Spinner flask

1. Add 200 ml of growth medium to a 1-l spinner flask, gas with 5% CO_2 and warm to 37°C.

2. Add cells from a logarithmically growing seed culture at 1-2 × 10 cells ml^{-1}.
3. Place spinner flask on stirrer and stir at 100-250 rpm (this is variable depending upon the cell type and geometry of the stirrer bar and vessel – a guideline is to use the minimum speed, which gives, by visual examination, complete homogeneity).
4. Monitor cell growth at least daily by taking a small sample from the side arm (remove flask to a tissue culture cabinet) and carrying out a cell count (Trypan blue stain and a haemocytometer).
5. Monitor the pH. In closed systems (i.e. all ports closed with no filters) the pH will become acidic and need adjusting by day 2-3. Remove the vessel to a tissue culture cabinet and: regas with 5% CO_2 in air; add sodium bicarbonate (5.5%); or allow cells to settle, remove around 50% of the medium and replace with fresh prewarmed medium.
6. After 3-5 days the culture density will typically reach 1-2 × 10^6 ml^{-1} (for many hybridomas, 0.8-1.5 × 10^6 ml^{-1} only) and can be harvested, or the culture prolonged for extra cell growth with twice-daily 50% medium changes. Suspension cells tend to have only a limited stationary phase and it is thus important to monitor growth more closely than in monolayer culture to ensure that healthy, rather than dying/dead, cells are harvested.

If cells are prone to clumping, or adhering to surfaces, then a medium with reduced Ca^{2+} and Mg^{2+} concentration should he used (e.g. suspension MEM). Additionally it is good practice to siliconize the glass vessels with Dow Corning 1107 (which has to be baked on, in an oven) or Repelcote (dimethyldichlorosilane).

There is a tendency to subculture suspension cells by dilution into fresh medium. However, some cells do not grow optimally under this regime and should be centrifuged (800 *g* for 5 min) and resuspended in new medium.

There is a wide range of spinner vessels available:

1. Conventional vessels from Bellco or Wheaton.
2. Conventional vessels but with large-bladed paddles attached to the magnet or base of the stirrer shaft. These vessels give more efficient stirring with better mixing at the lower speeds and are thus primarily aimed at microcarrier culture.
3. The Techne system uses a radial stirring action, which is designed for good mixing at low stirring speeds and to minimize mechanical stress to cells.

4. The Bellco dual overhead drive culture system uses a similar stirring system to the Techne and additionally permits continuous medium perfusion.
5. The Techne BR-06 Bioreactor uses a floating impeller, which again gives a gentle stirring action. An advantage of this system is that the vessel can be used equally effectively at various working volumes from 500 ml to 3 l. This allows an *in situ* build-up of cell seed by just adding extra medium as cell growth occurs.

If both side arms are fitted with filters, then continuous aeration through the head-space can be carried out, giving better control of pH and higher cell densities.

Suspension cells, e.g. hybridomas and BHK, have been the main consideration in this module because microcarrier culture is described in this chapter. However, most of the principles for suspension cells are applicable to microcarrier culture. The main difference is stirring speed. Microcarrier culture should be initiated at a very low speed (15-30 rpm) whilst cell attachment occurs (stir just fast enough to keep cells and carriers in suspension). Do not aim for homogeneity at this stage. In fact, it aids attachment to have all cells and carriers in the lower 30-50% portion of the medium. Once cell attachment is complete (3-8 h), increase stirring speed to achieve complete homogeneous mixing (40-80 rpm). As the culture develops, and the carriers get increasingly loaded with cells, the stirring speed can he increased further (70-120 rpm) to prevent excessive clumping of carriers. Also, the vessels with large paddles or the Techne stirrer must be used — a magnetic bar alone is insufficient.

Pilot-scale Suspension Culture of Hybridomas

The most common cell culture systems developed for pilot- and commercial-scale production of *monoclonal antibodies* (MAbs) are hollow-fibre and ceramic matrix modules, stirred bioreactors and airlift fermenters. These systems allow cultivation of cells in batch, fed-batch, continuous or perfusion mode. The selection of a culture system and culture mode for the large-scale production of a particular MAb should take into account the growth and antibody-production characteristics of the particular hybridoma line. This module therefore presents an overview of the important characteristics of these systems.

Pilot- and Large-scale In Vitro Systems for Hybridomas

Hybridomas can be cultured in either non-homogeneous or homogeneous cultivation systems. In non-homogeneous systems the cells are immobilized in a growth chamber. Most systems have growth chambers with hollow fibres or a porous ceramic matrix. In homogeneous systems cells are grown in bioreactors with either mechanical stirring or air sparging to keep cells in suspension. Short descriptions with lists of characteristics of these cultivation systems are outlined below.

Hollow-fibre and ceramic matrix modules

In hollow-fibre modules cells grow in the extracapillary space. The medium is pumped through capillaries and the pH and oxygen saturation are adjusted after each passage. The fibres are made of semi-permeable membranes, allowing diffusion of oxygen, low molecular weight nutrients and waste products but not the high molecular weight MAbs, which are harvested in a concentrated form from the extracapillary space.

The principle of ceramic matrix modules is the same as for hollow-fibre systems, with the exception that cells are not separated from the medium circulation by a membrane. Cells are held in the highly porous matrix and are supplied with nutrients by circulating medium through small squared channels in the matrix. Some characteristics of these two systems are:

1. High cell density of $>10^8$ ml^{-1}.
2. High product concentration – in hollow fibres: 400 mg l^{-1} HIgM, 1100 mg l^{-1} HIgG; in ceramic matrix modules: 300 mg l^{-1}.
3. Nutrient, oxygen and waste product gradients in the growth chamber.
4. Long diffusion distance for nutrients and oxygen in hollow-fibre modules.
5. Product harvesting with no, or few, cells.
6. Estimation of cell density in the growth chamber only via oxygen consumption.
7. No access to representative cell samples for viability and morphology check during cultivation.
8. No hydrodynamic shear forces in hollow-fibre modules
9. Hollow-fibre membrane may clog.
10. Scale-up limited because the tube length cannot exceed a certain length and only the diameter of the modules or number of installed modules can be increased.
11. Transfer of cells, grown in small systems, into large systems is difficult.

Stirred bioreactor

Different agitation systems, e.g. turbine-type or marine-type impeller, vibromixer and cell lift, were developed in order to prevent exposure of animal cells, cultured in stirred bioreactors, to high shear forces. In large-scale or high-cell-density perfusion cultures, air sparging or oxygen-permeable silicone/polypropylene tubing will provide adequate aeration.

1. Homogeneous cell suspension; even supply of substrate in bioreactor.
2. Controlled cell environment (pH, oxygen, nutrients).
3. Easy access to representative cell samples for viability/morphology check.
4. Easy transfer of cells to the next scale-up step
5. Oxygen supply is the most critical parameter in large bioreactors or high-cell-density cultures. Long oxygen-permeable silicone/ polypropylene tubing will provide bubble-free aeration. The tubing may need to be anything from 0.5 m to 2 m per unit of liquid volume, depending upon the oxygen requirements. Oxygen transfer by air or oxygen sparging can be used but may be detrimental for shear-sensitive hybridoma clones.
6. Cells are exposed to hydrodynamic shear forces.

Airlift bioreactor

Air or oxygen sparged into the bioreactor at the bottom drives circulation to keep cells in suspension, instead of a mechanical stirring device. To improve circulation and oxygen input and minimize cell damage, the height to diameter ratio is normally high in airlift fermenters.

1. No need for long silicone tubing for aeration.
2. Cell damage due to shear forces produced by disintegration of air bubbles.
3. Addition of non-ionic detergent may be necessary to prevent foaming, especially in media with a high protein content.
4. Cell circulation may he enhanced with a mechanical stirring device to minimize the necessary sparging rate at low cell densities to a level sufficient to maintain dissolved oxygen tension at the desired level.

Cultivation Modes

The following cultivation modes have been described for production of MAbs in *in vitro* large-scale systems: batch, fed-batch, chemostat and perfusion. Batch and fed-batch processes are the most common methods. More demanding are chemostat and perfusion modes, which allow a continuous production of MAbs. All methods are applicable in homogeneous and, with the exception of chemostat, non-homogeneous systems.

In non-homogeneous systems and homogeneous perfusion systems, cells remain in the growth chamber and culture supernatant is collected intermittently or continuously. The cell concentration is not affected by harvesting, in contrast to homogeneous systems where cells are harvested together with the antibody. In the following, characteristics of homogeneous systems are described.

Batch and fed-batch mode

Cells are inoculated, grown up to a desired cell or product concentration and harvested. The duration of batch culture depends on inoculation density, cell line and characteristics such as growth rate and antibody-production kinetics of the cell line. After every run the bioreactor must be cleaned and autoclaved for the next run.

In fed-batch mode, cells are fed either continuously or intermittently and parts of the cell suspension are withdrawn periodically.

1. Low cell density compared with perfusion or non-homogeneous systems (at the end, less than 5×10^6 cells ml^{-1}).
2. Product concentration normally less than 100 mg H, but up to 200 mg l^{-1} in batch and 500 mg l^{-1} in fed-hatch is possible.
3. Low demand for process control.
4. Variable substrate and (waste) product concentration.
5. Bad ratio of production time versus regeneration time in batch mode.

Chemostat

Cells are inoculated and grown to a certain density, as in the batch mode. Subsequently the medium is continually pumped into the bioreactor at a constant rate and the cell suspension containing antibodies is withdrawn at the same rate. The dilution rate can be varied within the range of zero (batch) and maximal growth rate; higher dilution rates lead to wash-out of the cells. Cell density increases until one or several substrates (e.g. glucose, nitrogen) or waste products (e.g. ammonia) limit cell growth and steady-state conditions are reached. Antibody production may be continued for weeks or even months.

1. Cell density approaches the final concentration reached in batch cultures (1-5 $\times 10^6$ cells ml^{-1}).

2. Product concentration up to 300 mg ml^{-1}.
3. Constant cultivation conditions
4. Optimization for antibody production but not for biomass production is possible.
5. Optimization of parameters under steady-state conditions with high accuracy.
6. Higher product yields per reactor volume and unit time.
7. Flow rate of medium and liquid level in bioreactor must be controlled.

Perfusion

Perfusion is initiated in the same way as chemostat culture but, instead of the cell suspension being withdrawn, the cells are continuously separated by filtration or sedimentation and recycled into the bioreactor. The increasing cell population is fed by increasing the flow rate of medium pumped into the bioreactor. Continuous harvesting of antibody containing supernatant via a filtration device increases proportionally. The two most widespread filtration systems are rotating filter cages, located within the bioreactor or in an external loop, and tangential flow filtration units.

1. High cell density ($>10^7$ ml^{-1}).
2. Product concentration up to 1000 mg l^{-1}.
3. High yield per unit time and reactor volume.
4. Better utilization of medium.
5. Smaller reactor volumes for the same amount of monoclonal antibody.
6. Sophisticated instrumentation is necessary to control the process.
7. Clogging of cell separation filters may limit culture times.

Pilot-scale Suspension Culture of Human Hybridoma

The optimization of culture parameters and the scale-up of a human heterohybridoma in a stirred bioreactor are described in this section. From the viewpoint of scale-up and handling, a stirred bioreactor is chosen as the most practical approach for industrial-scale production of *monoclonal antibodies* (MAbs). Furthermore, stirred bioreactors are very flexible with regard to the optimization of culture parameters, i.e. oxygen supply (bubble free, air sparging) and culture mode (e.g. batch, fed-batch, chemostat and perfusion). They also give easy access to cell samples at any time of culture, and keep cells homogeneously supplied with nutrients and oxygen.

Optimization of Culture Parameters and Scale-up

The optimization of culture parameters is an important step in scale-up. The growth rate, the viability and the antibody production of hybridoma cells largely depend on culture conditions. Optimization studies were carried out in a 1.5-l bioreactor, equipped with pH and pO_2 regulators, that provides reproducible culture conditions and minimizes medium requirements. The results were then applied to a 12-l bioreactor.

Media

1. *Low serum-containing medium*: Iscove's modified Dulbecco's medium (IMDM) supplemented with 2% foetal bovine serum and 5×10^{-5} M β-mercaptoethanol.
2. *Serum-free media*: evaluation of serum-free media for human hetero-hybridomas has been described.

Cell line

The heterohybridoma (human × human × mouse) 4-8KH15, used for scale-up, secretes human monoclonal IgM specific for *Pseudomonas aeruginosa* lipopolysaccharide of Habs serotype 4. The generation and characterization of human hybridomas is described by Lang *et al.* (1989, 1990a).

Spinner flasks

Spinner flasks (Bellco) of 500 ml volume are placed in an incubator aerated with air enriched with 5% CO_2. Agitation speed was set at 100 or 150 rpm.

KLF 2000 bioreactor

The bioreactor KLF 2000 with a capacity of 1.5 l was used to optimize growth and antibody production conditions of the hybridoma 4-8KH15. This flat-bottomed bioreactor is equipped with a marine impeller and a draft tube with diameters of 4.7 cm and 4.9 cm, respectively. The height to diameter ratio is 2:4. Temperature is regulated to 37°C with a 800-W heat finger. Dissolved oxygen tension and pH are measured on-line with heat-sterilizable electrodes. Cultures are aerated bubble free with air and O_2 to maintain the dissolved oxygen concentration at 40%, and the pH is adjusted with CO_2 and NaOH. The bioreactor is sterilizable *in situ* by heating water, filled into the bioreactor, to 120°C.

The flat-bottomed design of the KLF 2000 bioreactor affords a higher stirrer speed to prevent sedimentation of cells compared with the round-bottomed NLF 22 bioreactor. For cultivation of shear-sensitive cells, round-bottomed bioreactors may be necessary. To avoid high temperatures near the heat finger, a power control unit is used to restrict the power to the range 800-200 W but as low as possible to maintain a temperature of 37°C. In addition to this, heating is restricted to intervals of a few seconds only and the glass part of the bioreactor is insulated.

NLF 22 bioreactor

The 12-l NLF 22 bioreactor is equipped in the same way as the KLF 2000. The impeller and draft tube diameters are 7.9 cm and 10.9 cm, respectively. The bottom of the NLF 22 is spherically shaped. The height to diameter ratio is 1:95. The temperature is regulated through an integrated water-jacket. The NLF 22 bioreactor is steam sterilizable *in situ* by heating water, filled into the bioreactor, to 120°C. Cultures are aerated bubble free at low cell density, and at high cell density are intermittently sparged with air/O_2 to maintain the dissolved oxygen saturation at 40%. Bubble size and sparging rate were not measured.

Inoculum Preparation and Optimization of Parameters

1. Grow hybridoma cells in stationary cultures in T-flasks and passage every 2-3 days in a ratio of 1:3 to 1:10.
2. From T-flasks, transfer 100 ml of cell suspension to a 500-ml spinner flask and add 200 ml of fresh medium. Initial cell density is about 10^5 cells ml^{-1}. Two days later, add an additional 200 ml of fresh medium.
3. From the spinner flask, use 300-500 ml of cell suspension at a concentration of 3-6 × 10^5 cells ml^{-1} as inoculum for the 1.5-l KLF 2000 bioreactor for optimization experiments.
4. For cultures in the 12-l NLF fermenter, apply the same parameters as optimized for the 1.5-l KLF 2000 bioreactor. Only the stirrer speed, which is strongly dependent on the vessel geometry, has to be re-evaluated in the 12-l bioreactor.
5. Transfer 1500 ml of cell suspension containing 7-8 × 10^5 cells ml^{-1} from the KLF 2000 to the NLF 22 bioreactor. The dilution rate of spent to fresh medium should be 1:8 and the cell density after inoculation should be about 1 × 10^5 cells ml^{-1}.

Discussion

Background information

Perfusion culture results in the highest yield of MAb per reactor volume and time and the highest product concentration achievable in stirred bioreactors. Despite the need for sophisticated instrumentation to control the process, this mode seems to be the most favourable for large-scale antibody production.

Independent of the chosen culture method, cells are inoculated at low density into stirred bioreactors and initially grown in batch mode. inoculum density, the ratio of fresh to conditioned medium and stirrer speed are the first parameters to be optimized in batch cultures. Normally, confluent cell cultures are passaged by transferring a portion of the cell suspension into new medium at a certain ratio. Alternatively, cells can be centrifuged and resuspended in completely fresh medium. Most cells grow better in partially conditioned medium as compared with totally fresh medium and at a cell density below 1-2 $\times$ 10^5 cells ml^{-1}. Therefore, low dilution ratios and consecutively a high ratio of conditioned medium may he advantageous, e.g. when the culture system is changed from stationary T-flask to stirred bioreactor. On the other hand, high dilution ratios are necessary to minimize scale-up steps. The stirrer speed has to be optimized in order to prevent cell sedimentation and cell damage due to hydrodynamic shear forces.

Parameters such as pH, pO_2 and growth rate can be optimized in chemostat cultures. First, cells are grown to a steady state (normally 3-4 liquid volume changes) with a chosen set of parameters. Second, the parameter of interest is changed to a new value and cell density, viability and antibody secretion are monitored. With this strategy different sets of culture parameters can be defined. One set maintains a fast-growing cell population and a high viability during scale-up. Alternatively, culture parameters can be adjusted for a high cell density and a high specific antibody secretion in a production-scale bioreactor.

In perfusion cultures, the perfusion rate for a given cell density, the circulation rate in the cell-retaining filter and the cell recycling rate should be evaluated. A complete cell recycling rate may lead to steadily increasing cell concentration and accumulation of dead cells and cell debris in the bioreactor. To operate the perfusion culture in a. steady state, the recycling rate can be set to a value below 100% or filters with a large pore size should be used to prevent accumulation of cell debris.

Critical parameters

The oxygen supply of mammalian cell cultures is the most critical parameter. In high-density cultures or in large biorcactors, oxygen supply may be insufficient. Factors influencing oxygen transfer into the medium are the oxygen concentration in the inlet gas, the stirrer speed and the area of the gas/liquid interface. The area of the gas/liquid interface can be increased by air or oxygen sparging directly into the culture. Alternatively, bubble-free aeration by oxygen-permeable silicon or polypropylene tubing submerged in the medium may be used. The sparging method can affect cell viability and can cause foaming in serum-supplemented media with a high protein content. Oxygen-permeable tubing in the range of 0.5 ml^{-1} to more than 2 ml^{-1} culture volume seems to be impractical. Another critical parameter in stirred bioreactor cultures comprises shear forces produced by agitation or disintegration of air bubbles. The sensitivity to shear forces is clone specific and varies within a wide range. Different low-shear mixing systems have been developed and may avoid cell destruction. The detrimental effect on cells produced by air sparging can be reduced by the addition of serum, non-ionic detergent or increasing height to diameter ratio of the bioreactor.

Results

In stirred batch cultures after 3-4 days a maximum cell density of 7-10 $\times$ 10^5 cells ml^{-1} with a viability above 85% was achieved. Optimal stirrer speeds were 100-150, 300 and 100-150 rpm in the spinner flasks, the KLF 2000 bioreactor and the NLF 22 bioreactor, respectively. The high stirrer speed of 300 rpm in the flat-bottomed. KLF 2000 is necessary to prevent cell sedimentation. In the NLF 22, with a spherically shaped bottom, a reduced stirrer speed is sufficient. Accordingly, the impeller tip speed was 0.8 m s^{-1} in the KLF 200 bioreactor and 0.8-1.2 m s^{-1} in the NLF 22, values that were not detrimental to cells. Up to 40 μg ml^{-1} IgM was produced in batch cultures in 3-5 days. The cell-doubling time of about 25 h and the antibody production in stirred cultures were similar to those achieved in stationary cultures. The 4-8KH15 cells grow in serum-free medium under the same conditions as in low serum (2%)-containing media. Maximum cell density achieved in continuous culture was 1.5 $\times$ 10^6 cells ml^{-1} and viability was 80-90%. Without any change in medium composition, optimization of growth rate and pH in chemostat cultures resulted in a threefold increase of the specific IgM production rate, whereas IgM yield per reactor volume and time increased fourfold.

At cell densities exceeding 5 $\times$ 10^5 ml^{-1} in the NLF 22 bioreactor, the dissolved oxygen tension had to be maintained by intermittently sparging O_2 directly into the culture and raising the stirrer speed from 100 rpm (impeller tip speed 0.8 m s^{-1}) to 150 rpm (tip speed 1.2 m s^{-1}). No negative effects on cell growth and viability were observed.

Time considerations

A batch experiment takes 3-5 days to grow cells to confluency, up to 5 more days to investigate antibody production during decline phase and 1-2 days for cleaning and sterilizing. Optimization of parameters in chemostat experiments requires 5-12 days per selected value. The time required depends essentially on the dilution rate. Provided that the 12-l NLF 22 bioreactor is the final scale for production, 18-22 days are required to grow cells from the working cell bank to production scale. To reduce the time to start a new culture in the production bioreactor, it is important to maintain a culture at the preliminary stages.

CHEMOSTAT CULTURE

Determining the effect of individual parameters on cell physiology in batch cultures is complicated by the exposure of cells to a continuously changing environment. This is because a simple batch culture is a closed system in which cell growth is accompanied by the depletion of nutrients and the accumulation of end products. However, in an open system where nutrient consumption is balanced by input of medium, and cell and metabolite accumulation are balanced by removal of spent medium and cells, it is possible to achieve a steady state where cell and metabolite concentrations remain constant. This can be achieved by the use of a chemostat, as demonstrated independently by Monod (1950) and Novick and Szilard (1950). The theory of the chemostat is based upon the principle of Monod, which states that, at growth rates less than the maximum, the specific growth rate of a culture (μ) is determined by the concentration of a single growth-limiting substrate (s). Monod described the relationship between the specific growth rate and the limiting substrate concentration using an equation that takes a similar form to the Michaelis–Menten model of enzyme kinetics:

$$\mu = \mu_{max} \, [s/(s + K_s)]$$

The term μ_{max} represents the maximum specific growth rate that occurs when the growth-limiting substrate is in excess, i.e. when s is very large and $\mu = \mu_{max}$. The saturation constant (K_s) is equal to the substrate concentration at $\mu_{max}/2$. Both K_s and μ_{max} may be determined experimentally by Lineweaver–Burk-type analysis, plotting $1/\mu$, versus $1/s$ and extrapolating to obtain $-1/K_x$ at the x-axis intercept and $1/\mu_{max}$ at the y-axis intercept. Alternatively, a more accurate estimation of these parameters may be made by fitting the Monod model to the experimental data using suitable curve-fitting software.

The specific growth rate (μ) has units of reciprocal time (h^{-1}) and can be related to the cell-doubling time (t_d) by the following:

$$t_d \ln 2/\mu$$

In the chemostat the culture volume is kept constant by removing spent medium and cells at the same rate as the addition of fresh medium. The rate of dilution of the culture (D) is a function of the culture volume (V) and the flow rate (F), where $D = F/V$. The rate of change of cell concentration (dx/dt) is therefore a function of the specific growth rate of the cell (μx) and the dilution rate (Dx):

$$dx/dt = \mu x - Dx$$

Provided that the dilution rate is less than the maximum specific growth rate, the cells will continue to grow until the substrate concentration in the fermenter becomes limiting. At this point, further cell growth is determined by the rate at which substrate is added to the fermenter, and at a constant dilution rate a steady state is achieved. Because at steady state the cell concentration is constant (i.e. $dx/dt = 0$), it follows that $\mu = D$. Thus, at steady state the specific growth rate can be controlled by varying the dilution rate.

However, it is important to note that the specific growth rate is only equal to the dilution rate when all cells in the population are viable. Below 100% viability, the precise relationship between μ and D is more complex than the simple model described above, and a number of studies have shown that the actual specific growth rate deviates from D, particularly at low dilution rates. This results from an increase in the proportion of dead cells in the population, which in turn can be related to an increase in the specific rate of cell death (μ_d) at low D. The specific death rate at a particular dilution rate is related to the proportion of viable (x_v) and non-viable cells (x_d), where:

$$\mu_d = (x_d/x_v)$$

Taking into account the contribution from the specific death rate, the actual relationship between μ, and D becomes:

$$\mu = D + \mu_d$$

Whereas the specific growth rate is determined by the rate of addition of the growth-limiting substrate, the cell concentration in the fermenter is determined by the concentration of the limiting substrate in the feed medium. At steady state the cell yield (Y) on the growth-limiting substrate can be represented by:

$$Y = x/(s_R - s)$$

where s_R is the substrate concentration in the feed medium, s is the residual substrate concentration in the fermenter and x is the total cell concentration $(x_v + x_d)$.

At steady state, the concentrations of cells and metabolites are constant and thus calculation of specific utilization or production rates for each metabolite is relatively simple. For example, the specific rate of utilization of a nutrient q_n at steady state is given by:

$$q_n = D(n_R - n)/x_v$$

where n_R is the nutrient concentration in the feed medium and n is the nutrient concentration in the chemostat.

Similarly, the specific rate of production for a cell product (q_p) is given by:

$$q_p = D(p - p_R)/x_v$$

In summary, it can be seen that the chemostat enables the study of steady-state cell physiology at predetermined growth rates or biomass concentrations.

Equipment

Fermenter vessel

Most fermenter vessels suitable for suspension cell culture can be readily adapted for chemostat culture. A device for maintaining a constant culture volume will be required, the simplest and most reliable of which is a weir that allows the culture to overflow to a collection vessel. Alternatively, a dip tube that draws medium from the surface of the culture can be mounted in the headplate of the vessel and the medium pumped to the collection vessel. Most manufacturers supply modified tubes for chemostat operation of their fermenters.

Pumps

Peristaltic pumps are most convenient because they can be used in conjunction with sterile pump tubing (e.g. silicone or Viton). Constant wear of the tubing can lead to changes in the flow rate or tubing failure, so provision should be made to change the tubing regularly using sterile connectors. Alternatively, by using a long section of pump tubing that can be advanced through the pump head when necessary, excessive wear of the tubing can be avoided.

The flow rate should be checked regularly and this can be achieved by fitting a graduated pipette on the reservoir side of the pump. Medium can be drawn into the pipette using a syringe, and the flow rate calculated from the time taken for a set volume of medium to be withdrawn.

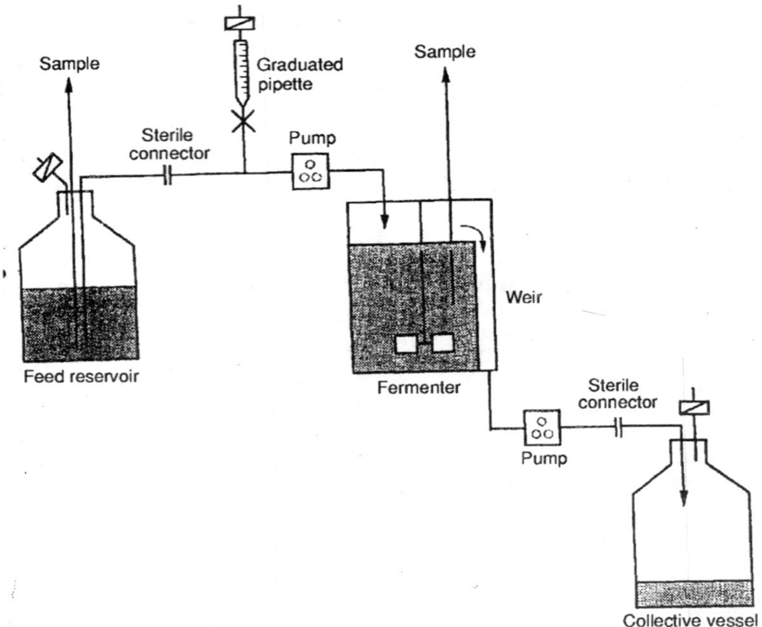

Fig. 2.2. Chemostat culture apparatus.

Media and collection vessels

The feed reservoir and collection vessel are connected to the inlet and outlet lines using sterile connectors so that they may be replaced as necessary. A sampling port should be fitted to the feed reservoir or at some point in the inlet line to enable sampling of the feed medium. The feed reservoir should be stored at 4°C to reduce degradation of labile medium components such as glutamine. If product is to be harvested, the collection vessel should also be refrigerated.

Method

Initiating a chemostat culture

Cells can be inoculated at a normal inoculum level, e.g. 1-2 × 10^5 cells ml^{-1}, and grown as a batch culture until the mid-exponential phase of growth. The supply of medium from the feed reservoir can then be started at a flow rate that gives the required dilution rate. In order that the cells are not washed out of the fermenter, the initial dilution rate should be below the maximum specific growth rate of the cells, although too low a dilution rate may result in low cell viability and oscillations in cell numbers. A dilution rate of between 50% and 75% of μ_{max} is usually a suitable starting point.

An indication of the value of μ_{max} for a particular cell line may be derived from batch culture experiments, and for most mammalian cell lines it is likely to fall in the range 0.02-0.06 h^{-1}. If the dilution rate exceeds μ_{max}, then cells will be washed out of the fermenter at a rate that is a function of the dilution rate and maximum specific growth rate. Indeed, the maximum specific growth rate can be determined experimentally by increasing D until wash-out occurs and then measuring the rate of cell wash-out. The maximum specific growth rate can then be determined using the following equation:

$$\mu_{max} = \frac{(\ln x_t - \ln x_0)}{t} + D$$

where x_0 and x_t are the respective cell concentrations at the start and end of the time interval and t is the time interval.

Growth-limiting substrate

The chemostat is a particularly powerful tool for examining the effect of different nutrient limitations on the physiology of mammalian cells. Selection of the growth-limiting substrate will depend on a number of factors, but most importantly it must be a nutrient essential for cell growth, e.g. essential amino acids, glucose, oxygen. The choice of substrate will also depend on cell type, because some nutrients will be essential for some cells but not for others. For example, glutamine can be regarded as an essential nutrient for most hybridomas because cells of lymphoblastoid origin lack glutamine synthetase and are unable to synthesize glutamine from glutamate. Conversely, Chinese hamster ovary cells express glut-amine synthetase and will switch to glutamine synthesis in response to glutamine limitation.

The concentration of the growth-limiting substrate in the feed medium should be set to that it becomes limiting while other nutrients are still present in relative excess. Small-scale batch culture experiments in which the cell yield is determined at different initial substrate concentrations will give some indication of a suitable substrate feed concentration (s_R). Once at steady state, it can be confirmed that the cells are limited by the chosen substrate because a change in the substrate feed concentration should give a proportionate change in the cell concentration.

Characterization of the steady state

Steady state is achieved when cell (as measured by cell number, biomass, DNA or protein content) and metabolite (e.g. glucose, lactate, ammonia, amino acids, product) concentrations remain stable over a period of time. The attainment of steady state should be confirmed by measuring several parameters, because the stability of a single parameter (e.g. cell number) is not necessarily indicative of a steady state.

The time taken for a culture to adjust to a new steady state will depend on the dilution rate, because it is advisable to allow five culture volume changes following a change. Thus it takes 100 h to reach steady state when D 0.05 h^{-1}, but 500 h when D 0.01 h^{-1}.

Sampling

Once at steady state, several samples should be collected over a period covering at least three culture volume changes, to ensure that the data generated are representative of the steady state. The volume of sample removed for analysis will depend on the number and type of analyses to be performed and the frequency of sampling. Removing large samplcs from the fennenter should be avoided, because this changes the volume of the culture and thus affects the dilution rate. Sample volumes of no more than 1-2% of the culture volume are recommended and the frequency of sampling should be reduced at low dilution rates. Samples should also be removed from the feed reservoir to check the composition of the feed medium.

Discussion

More extensive discussions of the theoretical aspects of chemostat culture have been published elsewhere. Many of the theoretical principles applied to microbial chemostat cultures can be applied successfully to animal cell cultures, although there are a number of instances where the behaviour of animal cells has been reported to deviate from the models used to describe microbial growth kinetics.

Some of these differences may be ascribed to the complex nutritional requirements of animal cells, which precludes the use of single carbon or nitrogen sources as limiting substrates. For this reason, the growth-limiting substrate is often not defined in animal cell chemostat studies, although a number of studies using defined growth-limiting substrates have been published. Glucose has been used as a limiting substrate in many chemostat studies on animal cells. Phosphate, oxygen-, and glutamine-limited chemostats have also been described in the literature.

Chemostat cultures have been employed in studies on the effect of the specific growth rate on mammalian cell physiology, gene expression and culture productivity. Chemostats have also been used to study the effect of specific nutrients and metabolites on cell physiology and protein glycosylation. The effects on cell growth of other parameters, such as pH, interferon, growth factors, Pluronic F-68 and gas sparging, have been investigated at steady state using chemostat cultures. Others have used chemostats with partial cell retention, which enables the study of cell behaviour at steady states that are difficult to achieve in a completely open chemostat system, e.g. high cell densities and low specific growth rates.

Growth of Human Diploid Fibroblasts for Vaccine Production Multiplate Culture

For many years human diploid cells have been utilized for the large-scale manufacture of viral vaccines. With the most recent developments in biotechnology, these cells are also capable of producing a variety of protein products. The history of establishment, growth and storage of human diploid cell lines has been extensively investigated and well documented. The absence of spontaneous transformation and adventitious viruses, a stable diploid karyotype and support of growth of a wide range of viruses are some of the reasons why these cells have been the substrate of choice for the production of biologicals. The growth of these cell lines has been greatly facilitated by the development of microcarriers. In microcarrier cultures, cells grow as monolayers on the surface of small spherical beads that are suspended in a suitable medium and in a vessel with constant stirring. The advantage of using this type of cell culture methodology is a homogeneous, well-controlled cellular environment.

Propagation and Subcultivation of Human Diploid Cells in 150-cm^2 Plastic Culture Vessels

This method describes the establishment and subcultivation of 150-cm^2 cell cultures originating from a human diploid working cell bank. The harvests of these cultures will be used to seed further similar vessels or microcarrier culture systems.

Materials and equipment

- Ampoule/vial of MRC-5/WI-38.
- Trypsin in phosphate-buffered saline (PBS).
- Dulbecco's modified Eagle's medium (DMEM) with 10% foetal bovine serum (FBS).
- 150-cm^2 plastic culture vessels.
- Harvest vessel.

1. Place frozen ampoule/vial of MRC-5/WI-38 from a working cell bank (3.0×10^6 cells at a predetermined population doubling level (PDL)) in warm water (37-40°C).
2. Shake gently until the contents are entirely thawed.
3. Remove thawed ampule/vial and swab with 70% ethanol.
4. Transfer contents aseptically, with a pipette and pro-pipetter, into a 150-cm² flask containing 100 ml of DMEM with 10% FBS.
5. Incubate at 37°C in a 5% CO_2/95% air incubator.
6. Fluid change the medium the following day and observe cultures macroscopically and microscopically during the incubation period.
7. Once the desired level of confluency is achieved (4-6 days, 25-30 $\times 10^6$ cells), transfer the culture(s) from the incubator to a tissue culture cabinet.
8. Pour off spent medium into a discard bottle, then wash cell sheet with 20 ml of PBS and discard immediately.
9. Add 10 ml of warm trypsin solution using a 10-ml plastic pipette and pipetter.
10. Gently roll the solution over the cell sheet and pour off the trypsin into the discard bottle.
11. Incubate culture(s) at 37°C until cells have loosened and separated.
12. Transfer the culture(s) from the incubator to the cabinet and break cells up with 10 ml of DMEM with 10% serum by pipetting up and down. Transfer the suspension to a cell harvest vessel.
13. Rinse culture flask(s) with an additional 10 ml of the cell-suspending medium and transfer the suspension to the harvest vessel. Pipette up and down to ensure homogeneity.
14. Remove 1 ml of cell suspension into a disposable culture tube for a cell count.
15. Determine the cell concentration of the suspension by direct counting and monitor the viability using the Trypan blue exclusion test.
16. If the following transfer is into 150-cm² flasks, seed a minimum of 3.0×10^6 viable cells per flask. Incubate at 37°C in a CO_2 environment (3-5 days).

Seeding, Cultivation, Trypsinization and Infection of Nunc 6000-cm² Multiplate Unit

The following method describes the seeding, cultivation, tiypsinization and inoculation of human diploid cells in Nunc 6000-cm² *multiplate* (MP) units. These MP units may be used to provide human diploid cells for vaccine production in essentially two formats: cells may be provided as monolayer cultures in the MP units for subsequent viral infection; or cells may be trypsinized out of the MP units and used to seed microcarrier culture systems or other MP vessels.

Materials and equipment

- MRC-5/WI-38 cells in exponential growth phase
- DMEM with 10% FBS
- Viral inoculum
- Viral growth medium
- Trypsin in PBS
- Nunc 6000-cm^2 multiplate unit with a 0.2-μm disk filter unit
- Cell harvest vessel with connection/air filter and magnetic bar
- Seeding vessel with connections, air filter and magnetic bar
- Trypsin vessel (500 ml) with connections and air filter
- Glass bell-end attachment
- Multiple-end connection
- Laboratory stand
- Cautery pump
- Spring or screw clamps

1. Remove MP unit from its plastic packing and inspect visually for cracks, faulty seals, etc.
2. Under sterile conditions in a tissue culture cabinet, remove the seal from one of the adapter caps and insert a sterile disk filter (0.2 μm) in the cap. Remove the seal from the second adapter cap and insert a 5/16-inch stainless-steel connector into the second cap. The stainless-steel connector is part of a multiplate-end connection designed to facilitate transfer of ingredients and cells, as well as final harvest. Each and every end of this connection contains sterile wrapping and should be clamped individually using spring or screw clamps.
3. For seeding, each unit requires 2-l of diluted cell suspension, prepared in a sterile 2-l Erlenmeyer flask equipped with a No. 10 rubber stopper, air filter and silicone tubing.
4. Secure the rubber stopper on the Erlenmeyer flask with a piece of tape. Then, aseptically connect the open end of the silicone tubing to one of the 5/16-inch stainless-steel connectors of the multiplate-end connection, and elevate the vessel on an overhead platform.

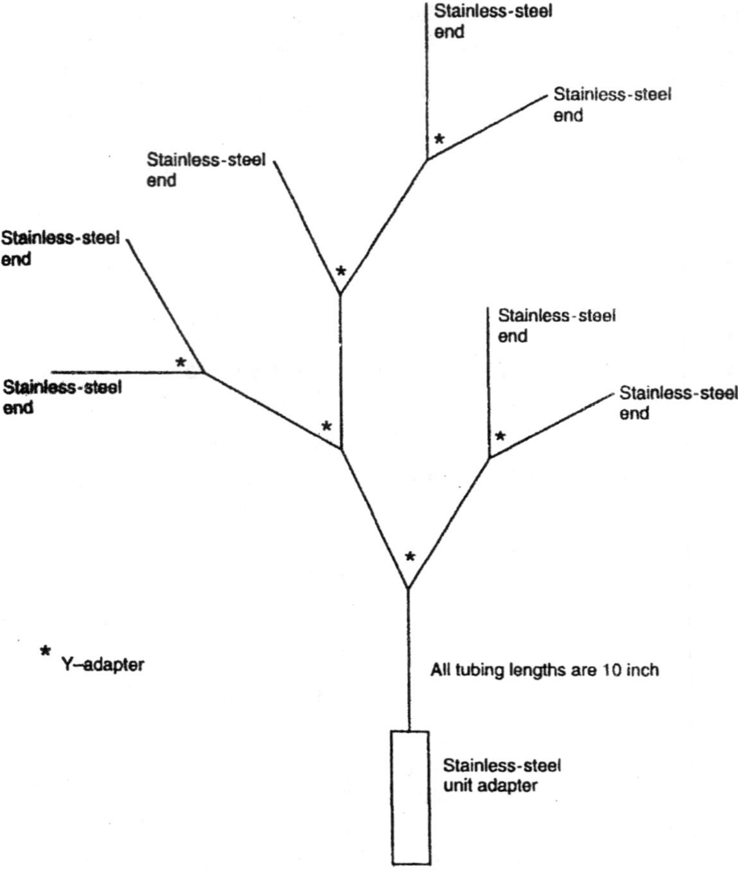

Fig. 2.3. Multiplate-end connection.

5. In order to facilitate the flow of all solutions by means of gravity, place the multiplate cell factory unit in a position with the supply tube to the bottom. To start the flow, a minimal amount of air pressure is applied on the unit using a cautery pump. Excessive air pressure may result in damage to the plastic seals.
6. Allow the cell suspension to flow in and ensure that an even level is achieved in each and every chamber. Clamp off the supply.
7. Rotate unit through 90° in the plane of the monolayer onto the short side away from the inlet.
8. Rotate unit through 90° perpendicular to the plane of the monolayer so that it now lies flat on its base with the culture surfaces horizontal.

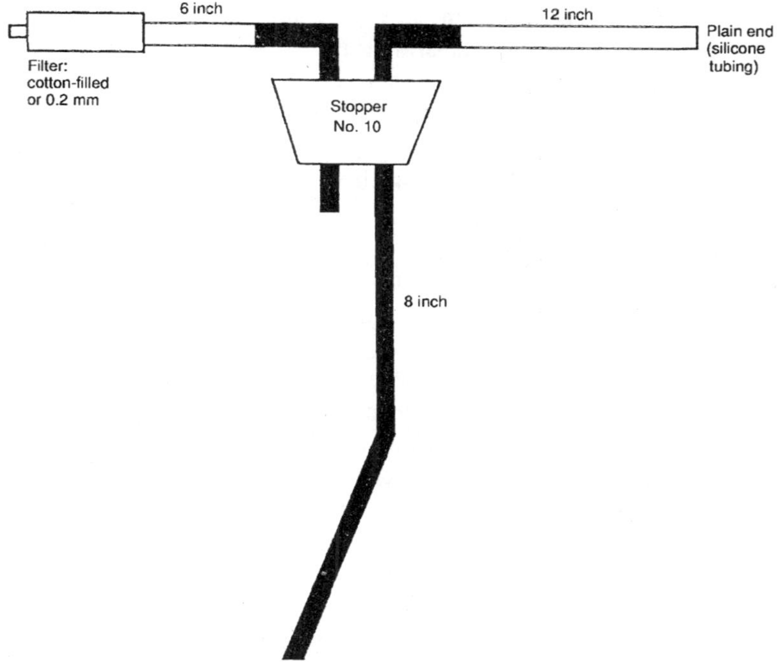

Fig. 2.4. Stopper connection plus one plain end.

9. Incubate MP unit culture for 3-5 days at 37°C in a 5% CO_2/95% air incubator.
10. When manipulating a multiplate cell factory unit, always lift from the sides or the bottom, never by the top plate.
11. Prior to harvest or viral infection, visually inspect the multiplate cell factory for any signs of contamination and/or leaks.
12. Connect four containers to the multiplate-end connection aseptically. The first vessel contains the trypsin solution and the second contains the growth medium required to dilute the cells and inactivate the trypsin. The third and fourth vessels are used to capture the spent medium (supernatant) and the cell suspension, respectively. The recommended procedure for connecting the vessels to the multiplate-end connection is described in detail in Steps 2-4.
13. Open the clamp of the supply tube and drain the spent growth medium by gravity. When the entire unit is empty, clamp off the portion of the tube leading into the discard vessel.
14. Fill the multiplate cell factory unit with 170-200 ml of trypsin and rinse the cell growth surface of each and every chamber.

Discard the excess trypsin, leaving a small volume that is just enough to provide superficial moisture. Clamp off the portion of the tube leading into the trypsin vessel. *Note*: Some operators prefer to wash the monolayers with PBS (Ca^{2+}-free and Mg^{2+}-free) prior to the addition of trypsin. When working with a large number of MP units, this additional step becomes impractical.

15. Incubate the multiplate cell factory unit at 37°C until the cells are observed to slide off the growth surface.
16. Once cell detachment is obtained, open the clamp of the tube leading from the vessel containing the growth medium and pump 300-500 ml of cell growth medium inside the cell factory unit. Rotate the unit so that the medium will dislodge as many cells as possible. Close the clamp of the cell growth medium vessel and open the clamp of the portion of the tubing leading into the cell harvest vessel. The cell harvest vessel is equipped with a magnetic bar. For better cell dispersion and homogeneous cell suspension, the magnetic bar must be activated as soon as the first volume of cell suspension reaches the harvest vessel.
17. Repeat Step 16 with an equal amount of medium, making sure that each and every chamber is properly rinsed.
18. Close off the tubing leading into the cell harvest vessel and let the cell suspension mix thoroughly for a few more minutes.
19. Connect a glass bell-end attachment to the multiplate-end connection. Pressurize the harvest vessel and, via the glass hell-end, remove a small volume of cell suspension for a viable cell count. An MP unit seeded with a minimum of 70×10^6 cells will yield approximately $0.8\text{-}1.6 \times 10^6$ human diploid cells. For operations involving viral infection instead of trypsinization, the following steps should be considered.
20. Proceed with Step 13.
21. Aseptically, connect the viral seeding vessel to the multiplate-end connection. The viral seeding vessel will contain the desired viral inoculum in viral growth medium.
22. Open the clamp of the tube leading from the viral seeding vessel and pump the contents into the MP unit. Clamp off and incubate the infected MP unit.
23. Following viral growth, proceed with downstream processing of supernatant.

Discussion

Background information

Human diploid cell lines have been utilized classically as *in vitro* hosts for the propagation of polio, mumps, rubella, cytomegalovirus, varicella-zoster, rabies, hepatitis A, respiratory syncytial virus, parainfluenza and many other viruses. Other uses in the biotechnology industry include large-scale cultivation for the production of various cellular products, such as human interferon beta.

For the production of viral vaccines, the cultivation of cells and virus is primarily carried out in a batch mode. Cells are expanded on a compatible growth surface to a desired confluency and then subcultivated repeatedly until a sufficient inoculum is achieved for large-scale cultivation. In order for the microcarrier cultivation to be an efficient process, critical inoculation and growth parameters need to be identified. Due to the limited *in vitro* lifespan of human diploid cells and for practical reasons of experimental reproducibility, a cell banking system is utilized. Each experimental run is initiated from cells frozen in liquid nitrogen, for a fixed number of subcultivations. In order to establish a small working cell bank, 3.0×10^6 cells per cryogenic vial (1×10^6 cells ml^{-1}) are frozen in DMEM containing 10% FBS and 7.5% dimethylsulphoxide, at a predetermined *population doubling level* (PDL) of 20-30. It has been established that microcarrier culture performance is largely influenced by the seeding inoculum density. For human diploid cells the maximum growth rate is achieved with lower cell concentrations. Conversely, high cell concentrations result in an extended lag phase and poor microcarrier culture growth. This phenomenon appears to be related to the synthesis of intrinsic growth inhibitors and is cell type dependent.

The use of Nunc cell factory technology for large-scale cell culture and viral vaccine production offers a variety of advantages. When utilized properly, the MP units require less than half the incubating and storage space compared to the T-flask or roller bottle system. There is also a reduced possibility for contamination. Once connected correctly, the units are a closed system and do not require multiple closures or openings. If a unit is found to be contaminated, it can be segregated easily from the remainder of the batch. An MP unit is processed in a fraction of the time required to process the 40×150 cm^2 T-flasks that it replaces For certain applications the units can be re-used, making the activity extremely cost-effective. However, the fact that the cell factory cannot be observed microscopically may he

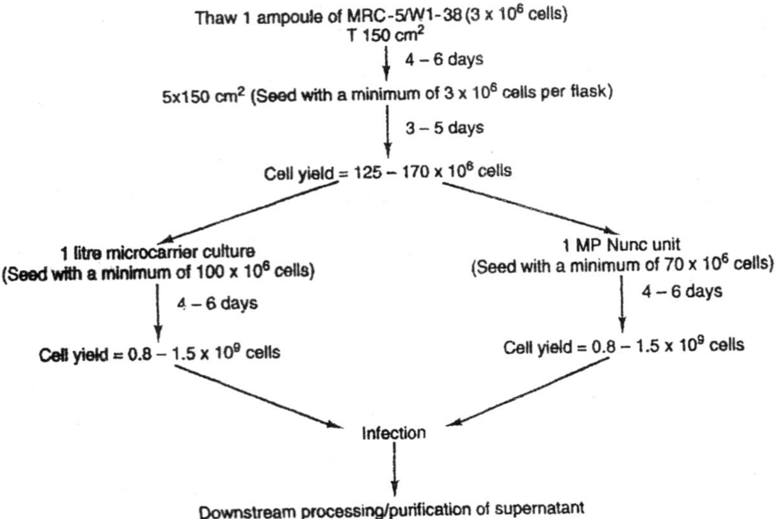

Fig. 2.5. The expected results and time frames when following a microcarrier or Nunc multiplate (MP) culture path.

the only drawback. This technical difficulty can be overcome by preparing a single plate unit in conjunction with the multiplate cell factory. By carefully examining the single plate, the operator can determine the growth pattern and cell morphology.

Troubleshooting

Suboptimal quality of the raw materials may be detrimental to the growth of human diploid cells. The most obvious component is serum. Sera other than FBS (newborn, calf and adult) appear to support the growth of MRC-5/W1-38, at least for the short term (two or three passages). Longer term passaging requires FBS supplementation, It is important to prescreen your serum, because not all FBS lots will support the growth of human diploid cells. Attention should also he given to the water purity and washing practices of the cultureware and to the substratum.

Microcarriers—Basic Techniques

Microcarrier culture, i.e. the growth of anchorage-dependent cells on small particles (usually spheres) 100-300 μM in size suspended in stirred culture medium, has made a tremendous impact on upstream processes. During the 1960s, the availability of *human diploid cell* (HDC) lines allowed a rapid expansion in the manufacture of human vaccines. However, large-scale production was restricted to using many replicate small cultures (flask and roller bottles) because HDC lines were anchorage-dependent, a restriction that did not apply to veterinary vaccines, which were being produced in large scalable fermenter processes (a suspension cell line, BHK, was licensed for the production of veterinary vaccines). The opportunity came when van Wezel (1967) showed that cells would grow on dextran beads in stirred bioreactors. However, the chromatography-grade dextran being used was unsuitable for consistent and reliable growth of cell lines, particularly HDC. After considerable developmental work by van Wezel and Pharmacia, a range of suitable microcarriers became available. The first industrial process using microcarriers was described by Meignier *et al.* (1980) for *foot and mouth disease virus* (FMDV) vaccine production. Subsequently a whole range of microcarriers based on gelatin, collagen, polystyrene, glass, cellulose, polyacrylamide and silica have been manufactured to meet all situations. The key criteria were to get the surface chemically and electrostatically correct for cell attachment, spreading and growth. To put this development in full perspective the following facts illustrate the sheer scale of opportunity that this method gives the cell culturist:

1. 1 g of Cytodex microcarrier has a surface area of 6000 cm^2 and, used at a modest concentration of 2 g l^{-1} gives 12,000 cm^2 l^{-1}. This is equivalent to 8 large or 15 small roller bottles.
2. Suspension culture systems, unlike bottles, etc., can be environmentally controlled and were optimized in the late 1960s to a minimum 100-1 capacity, i.e. a 1 × 100-1 fermenter was equivalent to 800-1500 roller bottles. The scale-up potential is 10,000 1 (for suspension cells), although currently only 4000 1 has been used for microcarrier culture.

The availability of microcarriers has not only opened up industrial opportunities but also allowed the laboratory worker to produce substantial quantities of cells and cell products for research and development purposes.

General principles

1. Cells differ in their attachment requirements, and thus a range of microcarriers should be assessed for suitability.
2. It is critical to have suitable culture equipment. Spinner flasks should have a special impeller because a magnetic bar is unsuitable. The shape of the vessel is important for homogeneous mixing, as is the stirring unit, which must operate smoothly without vibration at low revolutions (15-100 rpm). Culture vessels must be siliconized

to prevent microcarriers sticking to the glass, as well as the bottles used for preparing and storing microcarriers.

3. Microcarriers can be purchased as dry powders or sterilized solutions. The dry powders are swelled in Ca^{2+}/Mg^{2+}-free *phosphate-buffered saline* (PBS) (3 h), the supernatant is decanted and discarded and then the residual powder is washed and sterilized in the same buffer (50 ml g^{-1}) by autoclaving (15 lb in^{-2}, 15 min, 115°C). *Warning:* do not overheat. The powder microcarriers are 'softer' than glass and plastic and can be used in higher concentrations, thus giving vastly greater unit surface areas (e.g. Cytodex, 6000 cm^2 g^{-1}; glass/polystyrene, 300 cm^2 g^{-1}). Some microcarriers are available at different specific gravities (usually between 1.01 and 1.05), offering a choice depending upon the stirring (or mixing) system - roller bottles, shake flasks or airlift bioreactors can be used - and process requirements (e.g. if multiple medium changes are to be made, the heavier beads settle out more quickly and efficiently).
4. Consideration may have to be given to using a supplemented medium during the initial stages of culture to aid cell attachment and to offset the effects of low cell density, which will be more critical in microcarrier than stationary culture. The supplementation may be simple, e.g. non-essential amino acids, pyruvate (0.1 mg ml^{-1}), adenine (10 μg ml^{-1}), hypoxanthine (3 μg ml^{-1}) and thymidine (10 μg ml^{-1}). Other supplements include tryptose phosphate broth (1 mg ml^{-1}), HEPES (5 mM), transferrin (10 mg l^{-1}) and fibronectin (2 μg ml^{-1}). Serum, unless serum factors are added, may have to be used at 5-10% initially before being reduced after 1-2 days of culture.
5. The conventional cell-counting procedure of trypsinizing cells and then counting in a haemocytometer with Trypan blue can be tedious or even inaccurate (due to problems in removing cells from the microcarrier sludge). A better method is to use the nuclei-counting method developed by Sanford *et al.* (1951). The microcarriers are suspended in 0.1 M citric acid containing 0.1% crystal violet. The mixture can be vortex mixed or left at 37°C for at least 1 h. An advantage is that samples can be stored for long periods (at least 1 week) at 4°C before being counted. However, it is a total, not viable, cell count. Cells can also be fixed and stained on microcarriers for microscopic examination using standard procedures. Haematoxylin is the most widely used stain.

6. It is possible to re-use some microcarriers using good washing procedures and re-equilibration in PBS. Except for glass microcarriers this is not recommended, especially for more than one re-cycle, because performance drops significantly. There are reports of *in situ* re-colonization of microcarriers, which is feasible but again reduces the efficiency of the process (lower yields and heterogeneity in cell numbers between carriers).
7. Enzymic harvesting (e.g. trypsin) of some cells from microcarriers can be difficult or damaging. Consideration can be given to digesting some microcarriers (e.g. by dextranase, collagenase) because this gives a far quicker and less damaging means of getting a single-cell suspension.
8. Microcarrier culture requires more critical preparation than non-dynamic culture systems. To ensure success it is very important that all experimental details are carried out optimally. Of particular importance is the quality of the cell inoculum. This should be rapidly dividing, not stationary, and cells should be in a good condition (i.e. not trypsin-damaged) and as near a single-cell suspension as possible. It is good practice to feed the seed culture 24 h before use. Also ensure that medium is prewarmed and pre-equilibrated for pH (shifts in pH during attachment are extremely damaging). Due to the number of manipulations in the process, extreme care with aseptic technique should be taken, and as many as possible of the steps carried out in a tissue culture cabinet. Do not inoculate below the minimum number; the culture may not initiate or, if it does, will not reach maximum cell density.

Siliconization

Materials and equipment

- Dimethyldichlorosilane

1. Add a small volume of siliconizing fluid to clean glassware (spinner flask, bottles and pipettes used for handling microcarriers) and wet all surfaces.
2. Drain off excess fluid and allow glassware to dry.
3. Wash glassware in distilled water (combination of three washes/prolonged immersion).
4. Autoclave.

This siliconization process will allow glassware to be used through many repeat processes but glassware should be re-treated at least annually.

Growth of Cells on Microcarriers

Materials and equipment

- Eagle's *minimum essential medium* (MEM) (or equivalent alternative) with 10% *foetal bovine serum* (FBS), Eagle's non-essential amino acids and other supplements as necessary.
- Anchorage-dependent cell line (e.g. MRC-5)
- Microcarrier, e.g. Cytodex 3 (Pharmacia)
- Ca^{2+}/Mg^{2+}-free PBS
- Spinner vessel adapted for microcarrier culture.

1. Add complete medium to spinner flask (200 ml in 1-l flask), gas with 5% CO_2 and allow to equilibrate.
2. Decant PBS from sterilized stock solution of Cytodex 3 and replace with growth medium (1 g to 30-50 ml). Add Cytodex 3 to spinner vessel to give a final concentration of 2 g l^{-1} (use in range 1-3 g l^{-1}).
3. Put spinner on magnetic stirrer at 37°C and allow temperature and physiological conditions to equilibrate (minimum 1 h).
4. Add cell inoculum obtained by trypsinization of late log phase cells (pre-stationary phase) at over five cells per bead (optimum to ensure cells on all beads is seven); inoculate at the same density per square centimetre as with other culture types, e.g. $5\text{-}10 \times 10^4$ cm^{-2}. Cytodex 3 at 2 g l^{-1} inoculated at six cells per bead gives:

$$8 \times 10^6 \text{ microcarriers} = 4.8 \times 10^7 \text{ cells } l^{-1}$$
$$9500 \text{ cm}^2 = 5 \times 10^4 \text{ cells cm}^{-2}$$
$$200 \text{ ml} = 2.5 \times 10^5 \text{ cells ml}^{-1}$$

5. Place spinner flask on magnetic stirrer and stir at the minimum speed to ensure that all cells and carriers are in suspension (usually 20-30 rpm). It is advantageous if cells and carriers are limited to the lower 60-70% of the culture for this purpose. Alternatively either:
   (a) Inoculate in 50% of the final medium volume and add the rest of the medium after 4-8 h, or
   (b) Stir intermittently (for 1 min every 20 min) for the first 4-8 h. However, only use this option for cells with very poor plating efficiency because it causes clumping and uneven distribution of cells per bead
6. When the cells have attached (expect 90% attachment), the stirring speed can be increased to allow complete homogeneity (40-60 rpm).

7. Monitor progress of culture by taking 1-ml samples at least daily; observe microscopically and carry out cell (nuclei) counts.
8. As the cell density increases there is often a tendency for microcarriers to begin clumping. This can be avoided by increasing the stirring speed to 75-90 rpm.
9. After 3-4 days the culture will become acid. Remove the spinner flask and re-gas the headspace and/or add sodium bicarbonate (5.5% stock solution). With some cells, or at microcarrier densities of 3 $g\ l^{-1}$ or more, partial medium changes should be carried out. Allow culture to settle (5-10 min), siphon off at least 50% (usually 70%) of the medium and replace with prewarmed fresh medium (serum can be reduced or omitted at this stage). Replace spinner flask on stirrer.
10. After 4-5 days cells reach a maximum cell density (confluency) at the same level as in static cultures, although multilayering is not so prevalent. Thus a cell yield of $1\text{-}2 \times 10^6\ ml^{-1}$ ($2\text{-}4 \times 10^5\ cm^{-2}$) can be expected.
11. Cells can be harvested in the following way:
    (a) Allow culture to settle (10 min)
    (b) Decant off as much medium as possible (>90%)
    (c) Add warm Ca^{2+}/Mg^{2+}-free PBS (or EDTA in PBS) and mix
    (d) Allow culture to settle and decant off as much PBS as possible
    (e) Add 0.25% trypsin (30 ml) and stir at 75-100 rpm for 10-20 min at 37°C
    (f) Allow the beads to settle out (2 min). Either decant trypsin plus cells or pour mixture through a sterile, coarse-sinter, glass filter or a specially designed filter such as the Cellector.
    (g) Centrifuge cells (800 g for 5 min) and resuspend in fresh medium (with serum or trypsin inhibitor, e.g. soybean inhibitor at 0.5 $mg\ ml^{-1}$).

As a guide to calculating settled volume and medium entrapment by microcarriers, 1 g of Cytodex, for example, has a volume of 15-18 ml.

Discussion

The basic principles for using microcarrier culture are described, together with many notes on how to avoid problems and get the most out of this very powerful and useful technology. The description is suited to small-scale processes based on spinner flasks (i.e. 200 ml to 5 l) at levels that do not need special adaptations for perfusion, etc.

This does not mean that scale-up is not possible; in fact, it has been volumetrically scaled up to 4000 l for the production of interferon and viral vaccines. It has also been scaled up in density by the use of spin-filters to allow continuous perfusion of the culture and thus operation at microcarrier concentrations up to 15 g l^{-1} and cell densities over 10^7 ml^{-1}.

Porous Microcarrier and Fixed-bed Cultures

Porous microcarrier technology is currently the most successful scale-up method for high-density perfused cultures. The technology was pioneered by the Verax Corporation and systems are available from 16 ml to 24-l fluidized-bed bioreactors. The smallest system in the range, Verax System One, which is a benchtop continuous perfusion fluidized-bed reactor, was designed for process assessment and development. A great advantage is that the results achieved in the System One will scale up directly to the System 2000. The process is based on the immobilization of cells (both anchorage-dependent and suspension) in porous collagen micro-spheres. The spheres are weighted (specific gravity 1.6) so that they can be used in fluidized beds at high recycle flow rates (typically 75 cm min^{-1}). The micro-spheres have a sponge-like structure with a pore size of 20-40 μm and pore volume of 85%, allowing the immobilization of cells to high density (1-4 $\times$ 10^8 cells ml^{-1}).

The culture system is based on a fluidized-bed bioreactor containing the micro-spheres, through which the culture fluid flows upward at a

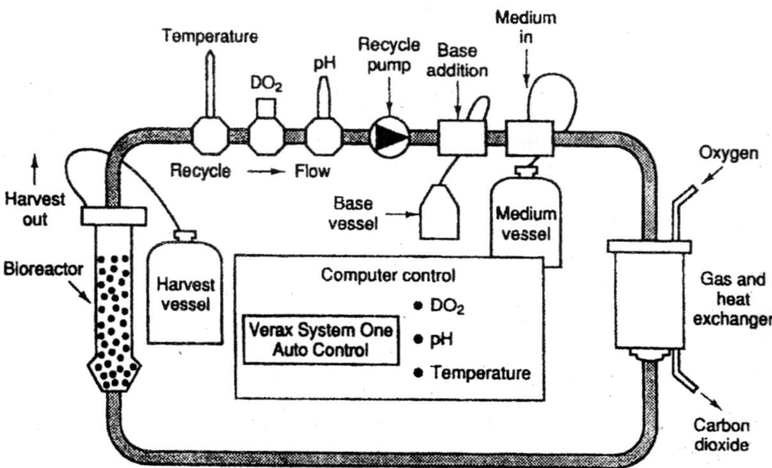

Fig. 2.6. Schematic diagram of Verax culture system.

velocity sufficient to suspend the microspheres in the form of a slurry. For oxygenation the medium is recycled through a membrane oxygenator. The system is run for long culture periods (typically over 100 days) by continuously removing the harvest and replacing it with fresh medium.

An alternative commercial system that is now available is the Cytopilot, which is a 25-l system using polyethylene carriers (Cytoline) that supports up to 1.2×10^8 CHO K1 cells ml^{-1} carrier.

Porous microcarriers can be used in stirred tank and fixed-bed cultures as well as the more commonly used fluidized beds. As stirred tank procedures are still very developmental, even though the design of specific porous microcarriers such as ImmobaSil is ensuring rapid progress, a procedure for fixed-bed culture is given in detail.

Fixed-bed Culture

The fixed-bed, porous-glass-sphere culture system was designed for the production of secreted cell products and lytic virus. The system is based on the immobilization of cells (anchorage-dependent or suspension) to high cell densities in porous glass spheres.

Large spheres (5 mm diameter) are used in fixed beds because they give an open bed structure (approximately 1 cm^2 channel cross-

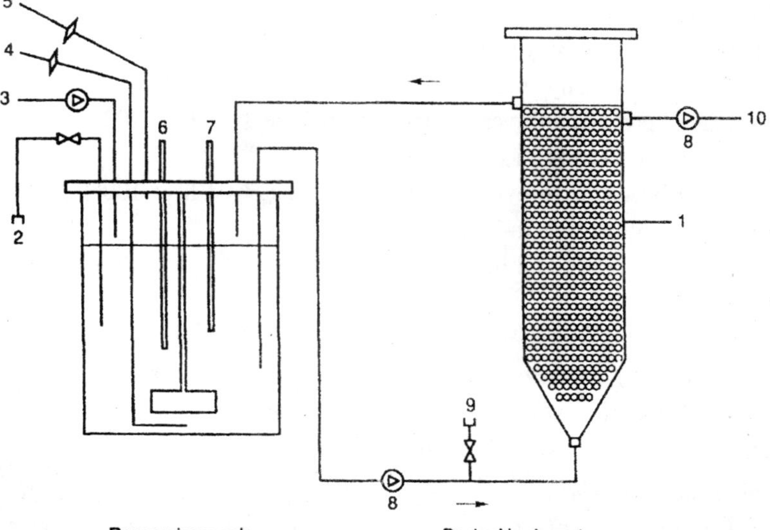

Fig. 2.7. Schematic diagram of fixed-bed porous-glass-sphere culture system: (1) packed bed of porous glass spheres; (2) sampling port; (3) medium fed in; (4) air/ oxygen sparge; (5) off-gas filters; (6) pH probe; (7) dissolved oxygen probe; (8) peristaltic pump; (9) inoculation port; (10) harvest.

sectional area), which minimizes blockages due to biomass build-up, uneven distribution of the inoculum and media channelling within the bed. The system consists of a reactor vessel containing the carriers for cell growth and a reservoir vessel for medium. A recycle flow of medium from 40 ml, increasing with cell growth to 450 ml (per litre packed bed volume per minute), is pumped up through the bed and returned to the reservoir for oxygenation. The system can be operated in batch, repeated batch feed and harvest, or continuous mode.

The reactor vessel is custom made from borosilicate glass with a water-jacket for temperature control. A 15-l (11-l working volume) Applikon stirred tank, or equivalent, reactor is used as a medium reservoir.

Initial Preparation and Calibration of Equipment

Materials and equipment

- 4% Decon
- 1.0 l reactor with water-jacket
- 15 l (11-l working volume) Applikon stirred tank or equivalent
- pH probes (Ingold)
- Polarographic dissolved oxygen (DO) probes (Ingold)

Bead preparation

1. Boil the beads in distilled water for 2 h.
2. Rinse thoroughly in distilled water.
3. Place beads in an oven to dry (approximately 4 h at 100°C).
4. Rinse thoroughly in distilled water.
5. Place beads in an oven to dry (approximately 4 h at 100°C).
6. Repeat Steps 4 and 5 three times.
7. Place dry beads in a beaker, cover with foil and seal with autoclave tape.
8. Sterilize the beads with dry heat at 180°C for 2 h.

Preparation of culture vessels

1. Wash the reservoir and reactor in 4% Decon.
2. Rinse thoroughly in distilled water.

Calibration of probes

1. Calibrate the pH probes (Ingold) according to manufacturer's instructions.
2. Calibrate the polarographic DO probes (Ingold) with N_2/air-saturated medium (at 37°C) according to manufacturer's instructions.

Assembly of Culture Vessels

Materials

- Silicon tubing
- Air filter and vent filter
- Marprene tubine
- Peristaltic pump
- Male and female stericonnectors
- Porous glass spheres (Schott Glaswerke)

Note: Maintenance of sterility during all steps involving additions or connections to sterilized equipment is essential.

Reservoir

1. Attach air filter with silicon tubing to sparge tube.
2. Attach air filter with silicon tubing to top air tube.
3. Attach vent filter with silicon tubing to air condenser.
4. Insert Ingold polarographic DO probe into DO holder.
5. Insert Ingold pH probe into pH holder.
6. Prepare the medium-out recycle tube, i.e. silicon tubing (5 mm diameter) with a short piece of marprene tubing inserted for the peristaltic pump, and a male stericonnector with blanking plug attached to the end of the tube.
7. Attach the medium-out recycle tube to the medium-out tube (long tube).
8. Attach short piece (50 cm) with a female stericonnector and blanking plug to the medium recycle return tube (short tube).
9. Attach three short pieces of silicon tubing with female stericonnectors and blanking plugs, i.e. for base and medium addition, with one spare entry port.
10. Place a Universal bottle on the sampling plot.

Reactor

1. Attach two pieces of silicon tubing with female stericonnectors and blanking pieces to Y-piece with single piece of tubing to inlet at base of reactor, i.e. for the medium-out recycle tube from reservoirs and inoculation vessel.
2. Attach silicon tube with male stericonnector and blanking piece to the medium-out recycle port on the reactor (cut to sufficient length to reach the medium-in recycle tube on the reservoir vessel).
3. Attach silicon tubing with male stericonnector and blanking plug to harvest port.

4. Insert probes (DO polarographic and pH) into probe holders.
5. Attach vent filter.

Sterilization

1. Add 200 ml of distilled water to reservoir and 100 ml of distilled water to culture vessel.
2. Cover stericonnectors with foil.
3. Autoclave at 121°C for 30 min.

Bead addition

1. Place the reactor in a tissue culture cabinet.
2. Spray the headplate with 70% ethanol.
3. Remove the headplate.
4. Remove the foil from beaker.
5. Pour beads into the reactor.
6. Replace the headplate.
7. Autoclave at 121°C for 30 min.

System Setup

Materials

- Thermocirculator
- pH controller
- DO controller

1. Attach thermocirculator with silicon tubing to reactor and reservoir, fill with water and set temperature to 37°C.
2. Connect medium recycle stericonnectors.
3. Place marprene tubing in the peristaltic pump head.
4. Add prewarined medium to the reservoir vessel.
5. Leave for 4 h to equilibrate.
6. Take a sample from the reservoir and measure pH using an independently calibrated pH meter.
7. Adjust pH controller to read the pH of the medium.
8. Check calibration of DO probe in reservoir according to manufacturer's instructions and adjust if necessary.
9. Adjust pH and DO controller set points to desired values.

Inoculation and Maintenance of Culture System

Inoculation

1. Resuspend inoculum (5.0×10^9 cells l^{-1} bed volume) in 800 ml (bed void volume) fresh medium.

2. Add to inoculation vessel.
3. Connect to reactor vessel.
4. Gently swirl the inoculation vessel to ensure an even cell suspension.
5. Slowly pump the inoculum into the bed with air pressure (hand pump), being careful not to introduce air bubbles into the system.
6. Drain and refill the bed twice.
7. For suspension cells, start perfusing immediately at 40 ml min^{-1} or at a linear flow velocity of 2 cm min^{-1}.
8. For anchorage-dependent cells, leave stationary for 2-4 h for the cells to attach before perfusing at 40 ml min^{-1}.

Batch operation

Run the culture until the maximum product concentration is reached.

Repeated feed and harvest operation

The reservoir medium volume (11 l) is changed when the glucose concentration in the medium falls below 2 mg ml^{-1}, i.e. daily in steady/ pseudo-steady-state culture.

Continuous operation

The dilution rate (medium feed rate) is adjusted to give a medium glucose concentration of approximately 2 mg l^{-1}.

Supplementary Procedure: Analysis of Consumption and Production Rates in the Fixed-bed Porous-Glass-Sphere Culture System

Take a 10-ml sample daily from the reservoir, and perform the following:

- Viable cell count (free cells)
- Glucose assay
- Lactate assay
- Product assay

Process calculations

Batch, repeated batch feed and harvest modes

Glucose consumption rate (GCR):

$$\mathrm{GCR} = \frac{(G_1 - G_2)}{(t_1 - t_2)} V$$

Lactate production rate (LPR):

$$\mathrm{LPR} = \frac{(L_1 - L_2)}{(t_1 - t_2)} V$$

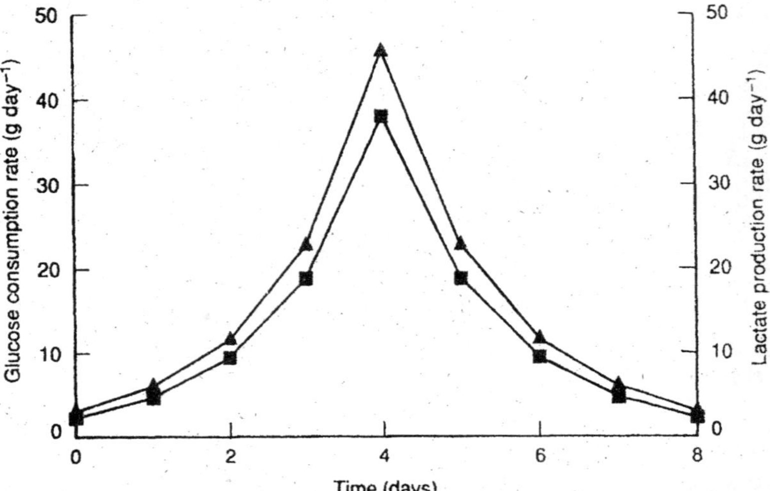

Fig. 2.8. Glucose consumption rate and lactate production rate in batch mode.

Product production rate (PPR):

$$\text{PPR} = \frac{(P_1 - P_2)}{(t_1 - t_2)} V$$

Continuous perfusion

Glucose consumption rate (GCR):

$$\text{GCR} = \text{MFR}\,(G_5 - G_3) - \frac{(\Delta C)}{(\Delta t)} V$$

when:

$$\frac{\Delta C}{\Delta t} = \frac{(G_3 - G_4)}{(t_2 - t_1)}$$

Lactate production rate (LPR):

$$\text{LPR} = \text{MFR}(L_2) + \frac{(\Delta L)}{(\Delta t)} V$$

when

$$\frac{\Delta L}{\Delta t} = \frac{(L_3 - L_1)}{(t_2 - t_1)}$$

Product production rate (PPR):

$$\text{PPR} = \text{MFR}(P_4) + \frac{(\Delta P)}{(\Delta t)} V$$

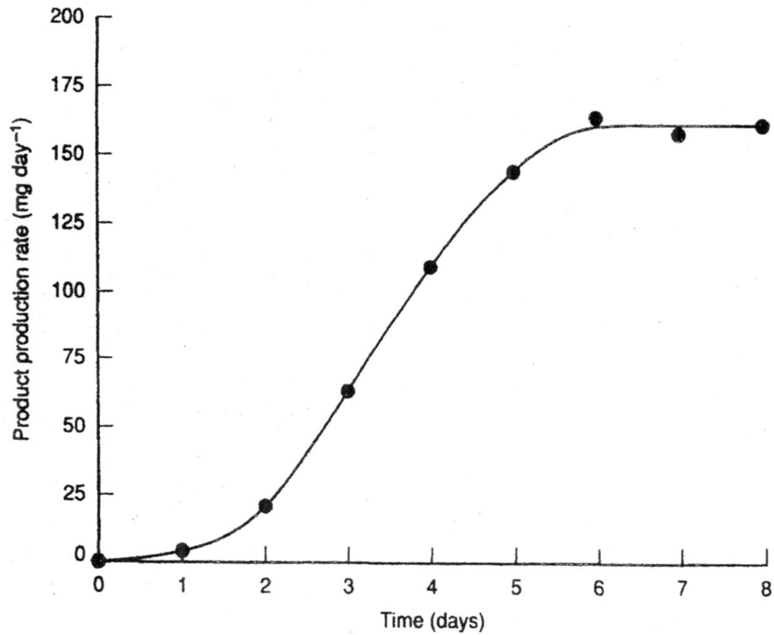

Fig. 2.9. Product production rate in batch mode.

when:

$$\frac{\Delta P}{\Delta t} = \frac{(P_3 - P_4)}{(t_2 - t_1)}$$

Example of fixed-bed culture system in different modes of operation

The medium used in these examples was Dulbecco's modified Eagle's medium (DMEM) with 4.5 g l^{-1} glucose, 4 mM glutamine and 5% foetal calf serum (FCS).

Termination of Culture and Determination of Cell Numbers

Run termination

1. Switch off DO controllers.
2. Switch off pH controllers.
3. Switch off thermocirculator.
4. Drain remaining medium/harvest from the system.
5. Drain water from the water-jacket.
6. Remove beads from the bed.
7. Take a 20-ml sample of beads.

8. Autoclave vessels and medium bottles.
9. Do a cell count on the 20-ml sample of beads.

Determination of final cell numbers (total)

1. Place 20-ml sample of beads into a 250-ml conical flask.
2. Add 100 ml of 0.1% citric acid and 0.2% Triton X-100.
3. Cover with Parafilm.
4. Place the flask in an orbital shaker at 37°C and shake at 150 rpm for 2 h.
5. Remove all supernatant and measure volume.
6. Wash the beads with phosphate-buffered saline (PBS).
7. Add crystal violet (10%) to the supernatant to give a final concentration of 0.1%.
8. Count the nuclei in a haemocytometer.

Discussion

Fluidized bed technology is ideally suited for the production of secreted cell products in long-term culture. It can also be used for lytic virus production; however, because this is a batch process, it is not ideally suited to fluidized beds. It must also be emphasized that this technology is not suitable for cell-associated products, e.g. some viruses.

A great deal of effort has gone into the development of high-density immobilized perfusion culture systems, which can be operated continuously. The limitation with many of these systems (e.g. hollow fibre) is that, whilst they achieve very high unit cell density (typically 10^8 ml^{-1}), they do not scale up well volumetrically. This limitation has been overcome by the use of porous carrier immobilization techniques, where high unit cell density can be combined with good volumetric scale-up potential and long-term continuous operation.

Fixed-bed solid-glass-sphere culture systems have been in existence for many years; however, they have not achieved widespread use, mainly because they have a low surface area per unit volume (0.7 m^2 l^{-1} for 5-mm spheres) and they are not suitable for the immobilization of suspension cells. These disadvantages were overcome by the introduction of the fixed-bed porous-glass-sphere (Siran) culture system, which has a very large surface area per unit volume (74 m^2 l^{-1}) and is suitable for suspension as well as anchorage-dependent cells. This has led to an increased interest in the potential of fixed-bed glass-sphere culture systems for the production of animal cell products.

Control Processes

The classical cell culture method (which is still used) is very simple but lacks efficient process control strategies. There is an incubator, providing reasonably good temperature control, and a CO_2-enriched atmosphere that interacts with a carbonate-buffered system to keep the pH within an acceptable range. All parameters for the regulation of the culture are taken off-line, mainly relying upon microscope observation by experienced operators. This strategy is appropriate for laboratory use, where a fully controlled, probably automated, process and good economy are not the main issues. For production processes, and for many investigative purposes where high reproducibility, high efficiency or high safety level, or all three, are needed, then more control and regulation of the environment is necessary.

Depending on the features needed, there is a wide range of possibilities available from commercial suppliers. The minimum configuration is a ready-to-use control cabinet measuring (and controlling) temperature, pH and pO_2 by set point devices. This may be sufficient for cell mass or product generation purposes using laboratory-scale, low-cell-density cultures. There should be no intention of creating a high-efficiency production process from such a set-up, because the very high cell densities are very sensitive to a balance between the physiological status of the cells and the environmental parameters. This cannot be achieved by set point control alone. Additional measurements and automatic, very quick and accurate control loops are needed for such processes, which are mainly dedicated to industrial production.

Defining the needs for the different levels of process sophistication is the first prerequisite for creating the appropriate set-up. To facilitate this, some of the principal features of processes and their control are discussed.

Basic Process Control

Process control for animal cell fermentation should at least include a set point control of temperature, pH, pO_2 and agitation rate.

Temperature control

As a consequence of low stirring rates, and the long mixing time inherent in animal cell fermentation, there is a high risk of localized overheating when using standard microbiological fermentation devices. Therefore, the best solution is a water-jacketed vessel fitted with warm water circulation, with the temperature limited to a value just 1-3°C higher than that of the reactor by a control loop overriding the reactor loop. Use hot water or low-energy electric heating elements rather than steam injection as the energy source to ensure minimum temperature variation.

Alternatively, for vessels up to 10 l, an electric heating band (barrel heater) may be used, having a power of approximately 30 W l^{-1} of reactor volume. In no case should heating rods or tubes inside the reactor be used, because localized high temperatures, aggravated by the low stirring rate, will cause unfavourable medium alterations and cell damage.

The control unit of the reactor should be a quick-response sensor inside the reactor (preferably Pt100 with a three-wire connection) and a proportional controller to achieve optimal temperature constancy. In no case should there be a deviation of more than 0.2°C during operation (during initial warming a little more has to be accepted).

pH control

The use of pressurized, sterilizable combined glass electrodes mounted through the reactor vessel wall is widely established. A high degree of accuracy and stability is needed to provide a good basis for the control loop. However, a recalibration is necessary both after sterilization and at regular intervals. This is easily done by external measurement after sampling the reactor, because the pH of the sample drifts very slowly.

There are system amplifiers/controllers supplied by the equipment manufacturers, but a standard laboratory titration instrument is also very useful for experimental fermentations.

With glass membrane pH probes there is some fouling of the membrane after several weeks of operation in protein-containing media. Therefore, for continuous processes that need accurate pH control, a changeable electrode mounting device should be considered.

Because all cell culture media are carbonate-buffered systems, the pH is dependent on the CO_2 in solution, which in turn is in equilibrium with the CO_2 in the gas phase. This offers the opportunity of using the CO_2 concentration within the gas phase for gentle, very efficient pH control. Alternatively, or in addition, pH control is possible by using base titration (acid titration is given by CO_2 anyway). When using hydroxide, one should use a 1 M mixture of NaOH and KOH (95:5) to avoid too great a shift in Na/K ratio and osmolarity.

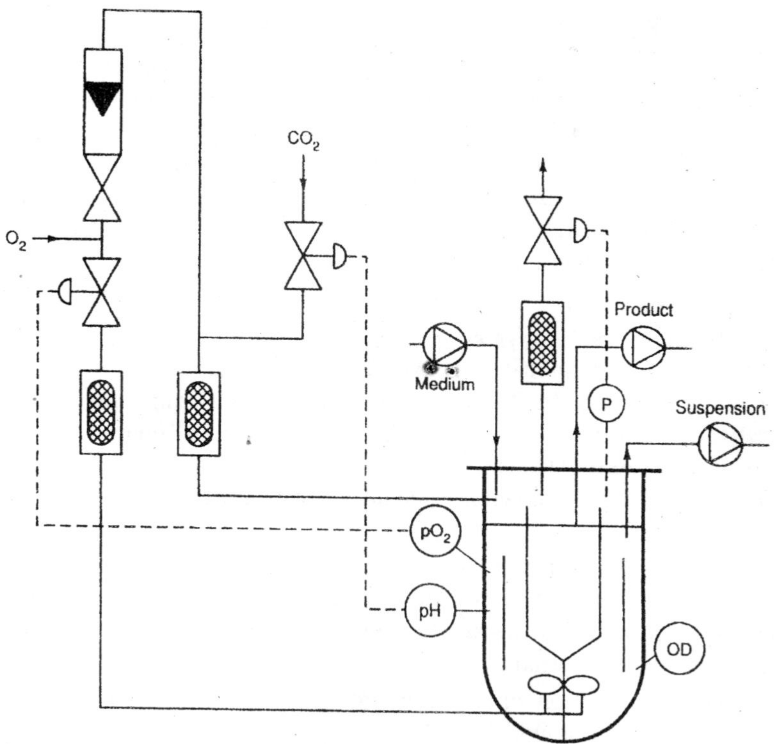

Fig. 2.10. pH control by CO_2, OD = optical density.

Control of dissolved oxygen

The availability of polarimetric electrodes means that there is now a stable and accurate sensor suitable for long-term animal cell fermentations. However, there are still several shortcomings:

1. The working cycle between the regular electrode services is limited to 6-8 weeks, depending on the oxygen tension. This could be overcome by the use of interchangeable housings or by the use of a second electrode, which is installed at the beginning of the run but connected electrically several weeks after the first electrode to save measuring capacity but to allow tuning while the first electrode is still in operation.
2. Recalibration after the start of the fermentation is difficult.
3. The response time for these electrodes is in the range of 1 min for 98% response.

To overcome these drawbacks with high-density cultures, a well-balanced design of the oxygenation system is necessary. The gas volume of the system from the control valve to the sparger outlet should be as low as possible to avoid over-titration. The proportional controller should have a self-optimization program that allows adaptation of the control characteristics to the electrode, pneumatics and sparger characteristics. Additionally, an overriding limitation of gas flow should be installed to minimize oscillations and restrict foaming. This can he done in a simple way by gas flow meters and needle valves or by using electronic mass flow measurement and control.

Control of agitation

'*Agitation rate*' is synonymous with '*stirrer speed*'. Because the stirrer speed with animal cell fermentations is much lower than for bacterial fermentations (20-200 rpm versus 1000-3000 rpm), there is a need for a slower drive, even in the case of a nominal control range starting from zero. This is because there is too much energy loss with wide-range electronic down-regulation, resulting in poor uniformity of speed.

With the proper mechanical drive range, all standard-speed controllers can be used successfully, but a tachometric control may be the most favourable solution. Additionally, this offers a convenient signal for speed recording.

The consideration of agitation rate/shear stress is a particular problem with scaling-up. Some systems show a favourable behaviour (e.g. airlift, circular loop), while others are quite difficult. For a detailed analysis of these problems, please consult the specialized literature.

Suppliers of standard cell culture fermentation equipment are:

- Applicon BV, Schiedam AC, the Netherlands
- Bioengineering AG, Wald, Switzerland
- B. Braun, Melsungen, Germany
- LH Fermentation. Ltd, Maidenhead, Berks, UK
- LSL Biolafitte, St Germain en Laye, France
- Marubishi Bioengineering Co., Ltd, Tokyo 101, Japan
- New Brunswick Scientific Co., Inc., Edison, NJ, USA
- Setric Genie Industriel, Toulouse, France

Enhanced Control of Physical/Chemical Parameters

The cultivation of cells at high densities and/or for long periods of time may be very attractive with respect to productivity and economy. Therefore, these methods will be used mainly for processes that are intended for industrial production. Such methods need not only a more detailed and safer process control, but also additional measures for meeting the requirements of good manufacturing practice.

Process control at high cell density is much more critical with regard to oxygen and nutrient supply as compared with standard batch and continuous cultures. The more intense use of medium contributes to the economy of the process, but leaves a deficit in nutritional buffer capacity. This causes rapid starvation of cells after relatively small alterations in specific nutrient supply. Therefore, cell number estimation, liquid flow/level control and the measurement of specific medium components and osmotic pressure will be dealt with here.

Cell number determination

Cell number and viability determination is not only an excellent direct process parameter, but also the basis for all specific and calculated parameters (growth rate, specific consumption and production rates). Therefore, the application of a sophisticated process control system is largely dependent on a reliable cell number measurement. Most laboratories determine the cell number within the reactor by sampling and off-line analysis. This is not satisfactory because of the necessity of frequent handling, overnight attendance and problems with reproducibility. On the other hand, automatic devices are either very complicated (like sampling photometers or sampling cell counters) or are relatively new and not validated.

The use of an infrared sensor may be a solution for many applications, particularly in combination with a control of viability. Other alternatives are sensors measuring conductance/capacitance and software sensors. A recently published method based on real-time imaging opens up new possibilities by real cell counting.

Liquid flow/level control

As outlined before, particularly with high-cell-density continuous processes, there is the need for accurate control of medium flow and liquid level. Wide variations in the feeding rate will lower viability, and variations in the liquid level will alter hydraulic behaviour and the oxygen transfer rate.

Liquid flow

The liquid flow cannot be controlled properly by peristaltic or other non-metering pumps because there are huge variations that are dependent on temperature, differential pressure, age of the tubing/ membranes, air bubbles, etc. This can be compensated for by the

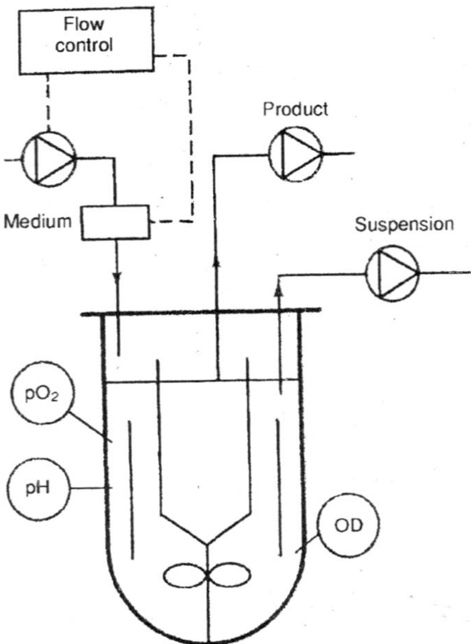

Fig. 2.11. Level control by electronic flow meter with a set point control loop.

introduction of an electronic flow meter together with a set point control loop. The flow may be measured using either magneto-inductive or thermoelectric principles. Very few companies can supply devices with a flow range as low as 1-5 l h^{-1}:

- Magneto-inductive principle: Turbo Messtechnik, Cologne, Germany
- Thermoelectric principle: Fluid Components, Inc., San Marcos, CA, USA

An alternative is provided by sterilizable metering pumps, but there are drawbacks: they are sensitive to gas bubbles and, for documentation purposes, an additional flow measurement device is necessary. A special device is made by Bioengineering Co., using a membrane valve combination for pumping (Kobio pump). A third possibility is the use of an automatic valve combination together with a small intermediate vessel that is placed on an electronic balance coupled to a computer for control of the valves. This is a very accurate method that can be adjusted to very small feeding steps down to a minute range, resembling an almost continuous medium supply.

Liquid level

Liquid level measurement may he accomplished in several ways. The classical way is the use of a liquid level sensor, either based on conductivity or on electrical capacity. They can be supplied by all fermenter companies. Both are affected by foam.

Another well-established method is to place the fermenter on an electronic balance. Most manufacturers offer such systems. They are very accurate as long as there are no changes with the peripheral installation (tubing, stoppers, etc.).

Two other useful methods should also be noted here:

1. Measurement of the differential hydrostatic pressure by two piezoelectric sensors mounted at the top and at the bottom of the reactor, respectively. The pressure difference corresponds to the liquid height.
2. Measurement by depth sounding using an ultrasonic device is also possible, but its use is dependent on a special mounting plate for the sonar head on the bottom of the fermenter.

Measurement of medium components

With high-efficiency processes it may be beneficial or even necessary to control the process by a distinct, critical medium parameter, either for adjusting the optimum feeding rate or for defining a distinct state of the process (harvest time, switching point, etc.). The parameter may be sugar level, concentration of a distinct amino acid, free ammonia, etc. As the metabolic behaviour of cell lines changes with many parameters, such as growth rate, cell density, pH, pO_2, glucose concentration and probably other physicochemical parameters, it may be useful to monitor more than one of the medium components. This will allow correct modification of the medium composition/ environmental conditions. The goal may be directed to cost-effectiveness or to favour the enhanced generation of a distinct isoproduct.

Measurement of osmotic pressure

The osmolarity of the cell suspension increases during a fermentation process as a result of metabolic events. The influence on the cells and the product is still unclear. Therefore, with each development of a new cell culture process, a study on the importance of this parameter must be performed. Unfortunately, there is no on-line measurement system commercially available, although sterile installation of a membrane osmometer should be possible. Currently, the standard method is off-line measurement using a freezing-point osmometer. For maintaining constant osmolarity, a second medium inlet system for distilled water must be installed and adjustments made occasionally according to sample readings.

Enhanced Control of Cell Metabolism

A rather more futuristic aspect of fermentation control is the use of complex parameters, calculated from basic and biochemical parameters. These include growth rate, uptake and production rates as well as intracellular metabolites. Such parameters directly monitor the physiological state of the cells and would allow very reliable and fast control of the process.

The automatic calculation of growth rate from two density measurements and the flow rate is not very useful because of the long time intervals necessary for a reliable calculation; this inherently leads to a retrospective value. A better possibility in this respect is the measurement of specific intracellular ATP, which correlates with the growth rate. This can be done by automatic off-vessel, on-line analysis of total ATP using commercial kits, the actual cell number from an on-line sensor and a calculation model. This value allows continuous fermentation at a constant growth rate by nutrient manipulation or temperature regulation. Productivity may be a function of growth rate, so this may lead to an improvement in process economy. Our own unpublished results show a correlation between glucose uptake rate and monoclonal antibody production rate with a hybridoma. However, due to limited technical resources, it has only become possible recently to use this phenomenon within a production process. Nevertheless, the automatic off-vessel, on-line measurement of glucose, twice per hour, can be established. An in-line glucose measurement system has been developed recently, based on near-infrared measurement.

A very interesting task would be automatic measurement of the product and the calculation of production rates. This would allow automatic optimization of the process by a computer program that varies all relevant parameters, probably by multiparameter analysis, to find the best production conditions. As long as there are no product sensors available, the main problem may be the time necessary for the measurement of an automatically taken sample. However, the use of HPLC methods can give accurate results within 20 min, and *high-performance capillary electrophoresis* (HPCE), with an analysis time of 5 min, could be introduced. Nonetheless, further development is necessary before these methods can be used routinely for automatic fermentation analysis.

Other parameters may include specific amino acid uptake or production rate, or the specific release rate of intracellular enzymes such as lactate dehydrogenase or glutamate oxaloacetate transaminase,

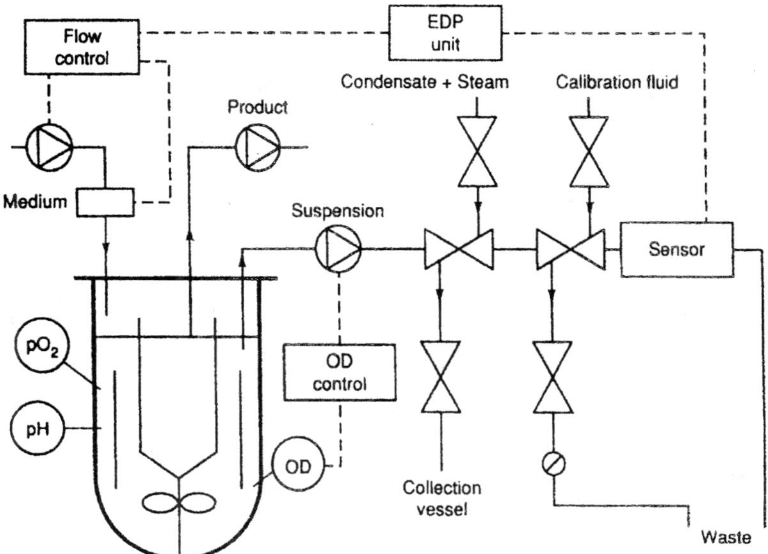

Fig. 2.12. Calculation of cell mass by on-line measurement of ATP, allowing a constant growth rate to be maintained. OD = optical density; EDP = electronic data processing.

both instant parameters for changing viability and particularly useful for alarm settings. Some of these possibilities are feasible today by using an automatic sampling device, either on a filtration base or free suspension sampling with an automatic analyser. For amino acids, a system called Biotronic LC5001 has been described. For enzymes, a modified clinical analyser system could be used; and for products (e.g. monoclonal antibodies), HPLC methods can be useful.

Discussion

In summary, it can be seen that basic process control may be satisfactory for laboratory fermentations, where the feasibility of the process should be investigated and a laboratory amount of material has to be generated without too much attention to economy. Enhanced process control will be necessary where either high product concentration or distinct product qualities are the goal and the manufacturing costs for the product are a major issue. Direct control on the chemical/physical level will probably give good results in many cases, but there is a clear advantage for process optimization in introducing control at the metabolic level. However, this may be a labour-intensive and costly task, which may pay back only with products achieving full production capacity for long periods of time.

3

MANIPULATING TISSUE

Tissue engineering (TE) is the application of engineering principles to cell culture for the purpose of constructing functional anatomical units, normally for reconstructive surgery. There are three main components: cell biology, materials science (polymer/protein biochemistry), and biological engineering.

TE aims to supply body parts for repair of damaged tissues and organs, without causing an immune response or infections, or using cadaveric tissue, or mutilating other parts of the patient. The dream is to blend the advantages of biological integration from living grafts with the ease, stability, and safety of prosthetic implants and as such it applies to most surgical specialities.

Two contrasting philosophies can be identified behind current approaches to tissue engineering. The first assumes that living cells possess an innate and self-sufficient potential for biological regeneration. The implication is that addition of suitable cell types to a suitable support matrix which allows proliferation and movement will result in an organized and functional tissue, resembling the tissue of origin. This is likely to be the simplest, most economic approach, where it can be applied. The second approach assumes that cells require a greater degree of control to produce new and functioning tissue structures. This approach reflects the understanding that, far from regenerating, mature tissues repair rather poorly *in vivo* and many cells do not organize in culture. This view leads to more complex solutions, generally supplying organizational cues to regulate resident cell function and provide spatial and synthetic control cues. Based on what is understood about tissue repair and regeneration biology it seems likely that epithelial and endothelial cell layers will prove better at organizing themselves through innate cell behaviours. In particular, cells such as keratinocytes and vascular endothelial cells make strong cell-cell interactions and form coherent sheets. On the other hand, connective tissues or mesenchymal layers, in which extensive deposition and remodelling of collagen matrix is needed (central to many TE applications), tend not to form appropriate structures spontaneously.

The contribution of tissue repair processes is important in TE developments, since these are almost all surgical implants, including skin, blood vessels, heart valves, nerves, joint surfaces, and ligaments. Consequently, there is a spectrum of stages at which TE constructs can be designed to be implanted. At one end of this spectrum is the engineered tissue which is designed to guide, control, and improve the repair function, effectively acting as an implanted provisional tissue, for example to support peripheral nerve repair. At the other end of the spectrum is a fully functioning tissue ready to work in the patient.

It is presently inconceivable that a functional peripheral nerve could be engineered to repair a defect or gap in a nerve tract. What is needed is a cellular repair tissue to guide and encourage regeneration of existing axons across the gap. This support function tends to be simpler to produce than a fully functioning implant, needing simpler and shorter *in vitro* tissue maturation periods, as this process occurs *in vivo*. However, tissue function will recover far more slowly and such constructs need to be functionally sophisticated to control local repair processes long after implantation.

Fully functioning implants include constructs designed to operate as large blood vessel implants or heart valves, where significant periods of patient recovery are impossible or impractical. Tissue applications falling between these two extremes can be tackled in a mixture of ways. Current forms of TE skin, for example, tend to resemble repair or granulation tissue more than a skin graft, although graftable dermis would have many advantages.

Design Stages for Tissue Engineering

A generalized scheme for the design of a TE construct with some specific examples. The surgical criteria are frequently based on previous procedures using grafts and prostheses, including ease of fixation, minimal patient discomfort, and the rapid restoration of function.

The source of donor cells is critical to the design. However, this is often overshadowed by the question of whether autologous cells are essential. Where a mature implant is needed and little remodelling is expected, use of the patient's own cells is favoured. However, allogeneic cells become more attractive where the TE construct is designed as a temporary repair tissue. The implanted cells have a temporary role and are expected to be gradually lost in remodelling. Allogeneic and xenogeneic cells can be immunogenic and need special precautions to screen for infection. However, they are more amenable for use in the mass production of consistent, rapidly available, low cost constructs than is the case where the patient's own cells must be used. In this respect the use of cultured cells appears to provide a fortunate advantage, in that prolonged culture or cryopreservation seems to reduce the antigenicity of allogeneic cells. Early experiences with new commercially available TE skin, using neonatal human foreskin cells, appear to support this view

Design criteria related to the support material include their porosity and structure. The survival time of the material is important, as are its degradation products, cell adhesion characteristics, and ability to propagate surface guidance and mechanical cues. These features will, by accident or design, control the types of local cells which are recruited (macrophages, polymorphs, fibroblasts, smooth muscle cells, epithelial, or endothelial cells), their spatial organization, and the nature of the matrix they assemble.

Three tissue examples are analysed (skin, urothelium, and peripheral nerve) in terms of their TE requirements and design features.

Tissue Engineered Skin

Tissue engineering of skin became feasible in 1975 with the demonstration that sheets of human keratinocytes could be grown in the laboratory in a suitable form for grafting. This was a simple, cohesive sheet of cells cultured from the donor on a feeder layer of fibroblasts. This technique has been extensively modified and applied clinically, but as a skin treatment without a dermal layer, its uses are limited. The epithelial component is able to regenerate in culture, since the cells grow as a continuous sheet over a suitable denuded surface, producing a continuous layer which progresses to form cornified layers. However, it is the underlying dermal layer which presents more difficulties, with its regular collagenous architecture, blood capillaries, nerves, and accessory organs such as sweat glands.

There are various forms of implantable skin substitutes which can be considered as TE constructs. The first and simplest is a basic collagen-glycosaminoglycan sponge known as *Integra*. Although Integra alone is a bioartificial material, rather than a TE construct, it has been used to cany seeded cells, as have a number of other collagen sponges and hyaluronan films. Integra consists of insoluble bovine collagen type I and the glycosaminoglycan chondroitin sulfate in a ratio of 98:2. *Dermagraft* consists of PGA polymer mesh of suitable pore size, seeded with human dermal fibroblasts from neonatal foreskins. As with Integra, this can be covered in a keratinocyte sheet at the time of implantation. *Apligraf* consists of human dermal fibroblasts seeded into a type I collagen gel and allowed to contract under tension. A layer of human keratinocytes is then seeded over the upper surface at the air-liquid interface. Both cell types in this construct are again derived from neonatal foreskins and so are allogeneic. Such TE constructs are available for surgical use, though with a limited shelf-life, presently in the order of five days. Clinical assessment of the performance and fate of Apligraf suggests that implants normally integrate well into surrounding tissues, forming a good skin cover. Importantly, there is no evidence of antibody formation to the bovine collagen, and little sign of rejection of the allogeneic cells in the construct, which are likely to disappear as the construct is remodelled.

Tissue Engineered Urothelium

Since human urothelial cells and bladder smooth muscle cells can be cultured, it is likely that construction of tissue engineered urothelial implants will be possible. The criteria are that the final structures need to form elastic tubes or bladders able to remain patent (i.e. without strictures), and the implant should not allow crystal formation from urine or harbour local infections. The structural requirements of the tissue are relatively simple in that an outer muscle layer should be lined on the lumenal surface by an intact, differentiating sheet of urothelium.

Support materials tested have included resorbable polymers [poly(glycolic acid) and poly(lactic-co-glycolic acid) co-polymer: PGA and PLGA] and cross-linked collagen sponges. Isolated urothelial cells cultured on collagen sponges formed differentiated sheets of urothelium over the surface of the material, with minimal tendency to promote crystal deposition. Urothelial and bladder muscle cells seeded onto PGA scaffolds formed urothelium-like, vascularized bilayered tissues when implanted into rabbits. Recentiy, this technique has been applied to the tissue engineering of a functional bladder in dogs using a fibrous PGA polymer base, shaped into a bladder, and coated in PLGA 50:50 co-polymer. Muscle cells were seeded onto the outer surface of the bladder after which the lumenal surface was coated with pre-cultured urothelial cells, prior to implantation. Implanted bladders achieved near-normal performance and maintained this for up to 11 months.

Tissue Engineered Peripheral Nerve Implants

Peripheral nerve injury is a common consequence of trauma and tumour resection surgery, with hundreds of thousands of reconstructive operations performed each year. Two criteria dominate approaches to assisted repair of peripheral nerve injuries. The first is that regeneration of axons should be guided as tightly as possible from their sprouting at the proximal stump to where they rejoin the degenerating distal stump on the far side of the defect. The second is that axonal regeneration across the nerve defect must be as fast as possible. Prolonged delay of reinnervation leads to irreversible muscle atrophy.

Nerve guidance was achieved using silicone conduits attached between the nerve stumps and also reported using tubes of bioresorbable materials such as PLGA co-polymer, collagen type I, and polymerized hyaluronan. The tube walls provide gross guidance, but no spatial cues are available to axons or Schwann cells away from the tube wall. Support for neurite outgrowth into the tube lumen has been provided by collagen and fibrin, with or without added growth factors, but again such gels provide minimal directional information.

Initial outgrowth for short distances from the proximal stump can be rapid, but for gaps of > 10-15 mm the process can slow or stop, probably due to a lack of supporting growth factors and Schwann cells. Continued growth can be achieved by adding purified growth factor or by seeding implants with Schwann cells. In another approach, optimization of the migration speed (as well as direction) has been described for orientated fibronectin implants, by varying the proportion

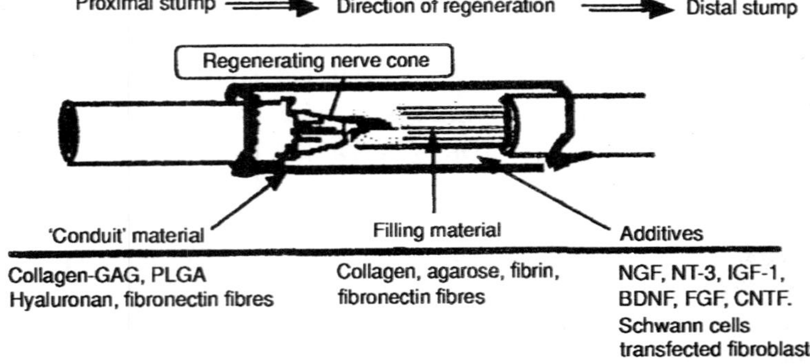

Fig. 3.1. Scheme illustrating the basic design and some variations for peripheral nerve implants. Three elements are commonly used.

of fibrinogen in the polymer material, thus altering its surface adhesion properties.

Cell Substrates and Support Materials

The value of a tissue engineering cell support depends on the information and suitability for the required adherent cell type or types. The information provided by the material is most commonly simple spatial cues, providing support surfaces with, for example, sufficient spacing for good cell growth. Biochemical information, such as surface reactive groups able to promote or reduce adherence or to activate cell membrane receptors (e.g. *integrins*) can be provided. More complex orientational (guiding cell responses, motion, and direction of deposition of extracellular matrix) and mechanical (tension, compression, or shear) information can be expressed through the supporting matrix of the construct.

The types of support materials available can be divided into discrete groups:

1. *Traditional*: abiotic materials; metals, plastic, ceramics.
2. *Bioprostheses*: natural materials modified to become biologically inert.
3. *Synthetic*: resorbable polymers.
4. *Semi-natural*: modified natural materials.
5. *Natural polymers*: proteins, polysaccharides.

Composite devices can be constructed using more than one of these materials.

For the purposes of tissue engineering constructs it is possible to largely omit the first group of traditional materials, since they are not designed to resorb or become biologically integrated within a reasonable period. Hence constructs based on these are more likely to be regarded as bioactive modifications of conventional prosthetic implants. In many cases, the same is also true of bioprosthetic materials in current use. These are materials formed by extensive chemical crosslinking of natural tissues, such as porcine heart valves and tendon. The native, collagen-based connective tissue is stabilized by glutaraldehyde treatment to produce non-immunogenic substrates which will survive largely unchanged at the implantation site for many years. Although cell layers and even new connective tissue can eventually grow over the surface of such bioprostheses *in vivo* they have been designed and fabricated primarily to function as long as possible independently and without modification by surrounding host tissues. The resistance to cellular infiltration and remodelling of bioprostheses is counter to the basic aims of tissue engineering and largely discounts their use.

A variety of synthetic bioresorbable materials are degraded by hydrolysis and then phagocytosed. The advantage of such materials is that production is relatively cheap and easy, in a controllable and reproducible manner at large scale. Disadvantages lie in their cell compatibility, which is often not as good as for natural polymers, and their degradation products, which can have unwanted cellular effects.

Polymer composition is critical. Varieties include PGA, PLA, polycarbonate, poly ε-caprolactone. The most widely applied polymers in tissue engineering are PGA and PLA and co-polymer PLGA. The composition, dimensions, and formation of these polymers can be adjusted to control their survival *in vivo* (stability), their gross mechanical properties (important in surgical handling and in replacing structural function), as well as their ability to support cell growth. In the case of PLGA this control is through adjustment of the ratios of PLA and PGA. PGA is a crystalline, hydrophilic polyester, typically losing its mechanical strength through hydrolysis over two to four weeks. In contrast, PLA is more amorphous and hydrophobic, degrading to release lactic acid and losing mechanical strength after eight weeks, though with much slower total resorption *in vivo*. PGA, PLA, and PLGA have been used to support cells in a range of tissue engineering models including cartilage, urothelium, smooth muscle, and skin.

The format of the support material is also important for different TE applications and relatively simple to modify using these polymers. The solid polymer screws and pins used in orthopaedics are less applicable to TE but early forms of polymer sutures were easily woven to give meshes and braids with porosities suitable for cell growth. Non-woven foams are made by incorporation of salt crystals into the polymer casting and subsequent dissolution by aqueous washing. Pore sizes are controlled by the size of the seeded crystals, typically between 150-300 μm. These materials can be moulded and extruded into a range of shapes including tubes for nerve guides.

Semi-natural and natural substrates are derived from natural macromolecular polymers or whole tissues. The distinction between them lies in the extent to which materials are modified (to achieve aggregation or stabilization) and how much this leads to frank denaturation. The most critical and pertinent test for TE applications is the extent to which a material participates in the natural remodelling process with local cells.

An example of a chemically crosslinked polysaccharide is mammalian hyaluronan, stabilized by benzyl esterification of increasing numbers of side chains. Hyaluronan is a charged polysaccharide (glycosaminoglycan) found naturally at cell surfaces and in the extracellular matrix. It is an important lubricant between gliding surfaces in soft tissues and has been implicated in a range of cellular functions, including angiogenesis and tissue repair. This aggregate, known as HYAFF (marketed as '*laser skin*'), is progressively more stable as crosslinking increases, but less of the material is biologically native and functional. Materials can be made with varying levels of substitution and these survive for progressively longer periods *in vivo*, eliciting only modest inflammation and local connective tissue response. HYAFF has been developed for use in skin grafting, particularly as a carrier for keratinocytes.

Collagen sponges are prepared from various forms of insoluble or aggregated collagen, acid extracted, and crosslinked with agents such as carbodiimide, tannic acid, or diphenylphosphorylazide. Some of these treatments produce substrates which can be utilized by cells *in vivo* and *in vitro*. A method for aggregating basement membrane into a cell substrate, termed TypeIV/TypeIV$_{ox}$, has been described using oxidatively crosslinked collagen purified from human placenta. It was designed principally to support growth of epithelial and endothelial cell layers and tested for the repair of skin and dura.

The most natural polymer materials are those in which protein stabilization is produced by drastic dehydration. This maximizes intermolecular interaction within polymers through complete removal of hydration shells, to promote intimate molecular packing. The most widely studied of these materials is the collagen-chondroitin sulfate aggregate material of Yannas and Burke, which is available as '*Integra*'. A range of forms have been tested with different added matrix components.

Aggregates of plasma fibronectin, produced as aligned fibrous materials, are derived from native protein solutions under directional shear and stabilized by dehydration. Their stability and attachment properties can be altered by treatment with trace levels of copper salts. For example, trace levels of copper differentially regulated growth and migration of Schwann cells and fibroblasts. This raises the possibility of producing modified substrates which have cell selectivity, potentially useful in segregating cell types between anatomical zones or layers.

The most natural forms of cell support are polymers whose aggregation can be achieved in culture as it occurs *in vivo*. Examples include Matrigel, fibrin glue, collagen gels, and some polysaccharides. Of the polysaccharides, chitosan and hyaluronan have been used in the form of a hydrated gel. Agarose gels are used to support a chondrogenic phenotype in chondrocytes. Matrigel is derived from tumour cells as a thick extracellular matrix, containing type IV collagen, laminin, proteoglycans, and growth factors. Despite being a natural substrate, its complex composition and tumour cell origin are limitations.

There are commercially available forms of fibrin glue or fibrin sealant designed for surgical use. These are prepared from fibrinogen and thrombin from human plasma, stabilized by a bovine protease inhibitor, aprotinin. One composite form, Neuroplast, comprising elastin and fibrin, has been proposed as a substitute for dura mater and tympanic membranes. Fibrin materials have also been used as implantable growth factor depots and agents promoting angiogenesis, forming a simple slow release depot for trapped agents. Growth factor depot formulation in fibrin stabilizes FGF activity.

Collagen gels or '*lattices*' were developed to study contraction as the gel shrinks or tension is generated by fibroblasts. Untethered lattices shrink, producing smaller and denser matrices, sometimes regarded as tissuelike materials, whilst tethered lattices generate substantial endogenous forces. Collagen gels have been seeded with other cell types, such as keratinocytes and endothelial cells. Numerous model tissues have been attempted, based on contracted collagen gelsas bioartificial grafts for skin replacement. The most advanced application is a bilayered tissue engineered allograft, '*Apligraf*' based on a fibroblast-contracted lattice and keratinocyte layer. Orientated forms of collagen lattice have been developed to provide spatial information to cells either mechanically or more deliberately, using magnetic fields, to provide contact guidance cues.

Cell Sources

The first requirement for cell culture in any TE application is to generate sufficient cells to seed. The stage at which cells are to be introduced into the implant, the amount of remodelling of the implant substrate required, the degree of function, and the timing are critical.

One of the most important questions which must be addressed is whether the source of cells should be autologous, allogeneic, or xenogeneic. The use of the patient's cells (autologous transplant) is often the most straightforward option. A biopsy of the appropriate tissue is taken from the patient, enzymatically digested or explant cultured, and the cells grown to the required numbers. Whilst this method has the advantage of avoiding the immune response and the possible transfer of infection inherent in the use of allogeneic cells, it does have its drawbacks. Depending on the type of tissue required and the condition of the patient (limiting factors here include disease state and age of patient), it may not be possible to obtain adequate biopsy material.

The cell type may pose problems, for example articular cartilage, which has a relatively low density of cells with a low mitotic capacity. Cosmetic issues influence the choice of skin for tissue engineering. Studies using keratinocytes derived from the sole of the foot indicate that the implanted cells resume their original phenotype and continue to produce a thickened epidermis. Genetically altering cells may alleviate this problem, but more needs to be known about the origins and maintenance of these regional cell differences within tissues.

Allogeneic sources of cells have advantages over autologous cells. It can be relatively easy to obtain healthy donor tissue, the cells may be cultured on a large scale at central '*factory*' sites, with none of the time constraints of autologous cells. The product will be cheaper, of a consistent quality, and available as and when required by the patient. The major problem associated with allogeneic cells is graft rejection. The level of immune response generated by cells differs between cell types. For example, endothelial cells can induce a large reaction, whilst fibroblasts, keratinocytes, and smooth muscle cells are less immunogenic. The reaction may also be dependent upon the donor age of the cells used, as fetal or neonatal cells can elicit little or no immune response. There have been many studies reporting methods for reducing the immunogenicity of transplanted cells, though with limited success. The current strategies involve segregation of the donor cells from the host using cell encapsulation techniques which allow only the passage in of oxygen and metabolites and the exit of cell-generated hormones. There must also be scrupulous staged analyses of cells for possible infections, such as hepatitis, HIV, and CJD, as well as screening for potential hazards, such as abnormal karyotype, or tumorigenic capacity.

Gene therapy has a role to play in tissue engineering. Cell populations can be altered by adding genes either to increase their output of a certain protein or add to their repertoire of expressed genes. Examples include the alteration of keratinocytes to produce transglutaminase-1, which is deficient in patients with lamellar icthyosis, a disfiguring dermal disorder. The engineered cells were transplanted in athymic mice and produced a normal epidermis. The technique has also been used to induce fibroblasts and keratinocytes to release factors they would not normally produce. Fibroblasts have been made to produce proteins such as transferrin, factor IX, and factor VIII. In the vascular system modified endothelial cells have been used to produce anti-thrombogenic factors such as tissue plasminogen activator. Chondrocytes have been engineered to produce an antagonist for IL-1, a major cause of degradation of cartilage ECM.

The method of culture of cells for TE implants will depend on the function of the cells. For many cell types, the classical monolayer (or equivalent large scale culture in roller bottles or spinners—may allow proliferation of cells, but also increases the possibility of phenotypic changes. Many cells revert to a fibroblast-like morphology after long-term monolayer culture (e.g. chondrocytes) or if allowed to become too confluent may lose their proliferative capacity (e.g. endothelial cells). This change may result in a loss of function due to the inability to produce, for example, a certain hormone or the capacity to contract. It must be assessed whether this alteration is temporary, i.e. cells revert to original phenotype when replaced in a 3D format or under the influence of soluble factors *in vivo*, or if the changes are irreversible. In either case it may be preferable to culture cells in an environment similar to the *in vivo* situation.

Perfusion is a critical issue in 3D culture. A number of factors interact to determine when nutrient and gaseous perfusion of a cell mass become limiting, including:

1. Cell density,
2. Metabolic rate,
3. Mean effective diffusion distance (cell to mixed nutrient source),
4. Density and anisotropy of diffusion path material.

Culture medium is exhausted more quickly by near-confluent cultures whilst some cell types, for example chondrocytes, survive at nutrient levels which would be critical to other cell types. In many experimental models, cells (e.g. *fibroblasts* or *chondrocytes*) grow in low density 3D aqueous gels. Collagen gels typically start at 95-98% water and these need to contract to less than 20% of their starting volume before they approach the cell and matrix densities of native tissues. Diffusion into such low density gels is rapid. In addition, 3D collagen lattices have a random, homogeneous fibril structure almost never encountered in native collagenous tissues. The native tissues have organized and aligned fibrous structures which will produce pronounced anisotropic diffusion properties. In particular, diffusion will be far more rapid parallel to rather than perpendicular to aligned collagen fibre bundles. Since matrix density and alignment are characteristics of mature rather than repair tissues, perfusion will be a greater issue in TE of mature tissues. The principal means of addressing this problem to date has been to incorporate a form of external pumping to generate a fluid flow around or through the tissue construct.

Throughout this section it is assumed that adequate perfusion of the construct will also optimize the rate of growth and the deposition of ECM. Control cues can also be optimized by adding growth factors and cytokines. Examples are the use of mitogenic growth factors to promote fibroblast proliferation (PDGF and FGF) or VEGF or FGF to optimize endothelial cell proliferation. Similarly, rates of matrix production may be regulated by 'control cues' such as the TGFβ and IGF families of growth factor. However, the incorporation of specific growth factors is extremely cell system- and stage-specific and this is made more complex by the need for multi-component cocktails.

Orientation

Cells require specific shape, directional, and spatial cues from their environment, including:

1. Contact or substrate guidance,
2. Chemical gradients, and
3. Mechanical cues.

In its simplest form, contact guidance uses the topographical features of the substrate. Topographical features are most frequently in the form of aligned fibres or ridges of appropriate dimensions (normally $< 100\ \mu m$ for fibres). Cell types which are aligned in this way include fibroblasts, tenocytes, neurites, macrophages, and Schwann cells. Bioresorbable guidance templates suitable for tissue engineering are based on synthetic polymer substrates, collagen fibrils, and fibronectin. Alignment of collagen fibrils within native collagen gels has utilized mechanical or magnetic forces, whilst tethering points, holes, or defects in the gel and the local effects of contractile cells can produce local orientation. Orientation of fibronectin fibres involves application of directional shear as part of the fibre aggregation process. A form of orientated fibronectin cable with a surface layer of aligned dermal fibroblasts, which migrate rapidly along the structure. It is possible to use differential attachment to substrates as a means of producing alignment in cells.

The speed of cell locomotion is cell type- and substrate-dependent, as demonstrated by the ability to alter cell speed on more or less adhesive substrates. Fibroblast speeds of 20-40 μm/h are common. It is possible to optimize cell speeds by suitable design of the substrate composition, but a far greater impact is made on recruitment by optimizing cell velocity. Random direction of movement will have a minimal velocity until some guiding cue is provided. In the case of cell guidance substrates, such as fibronectin fibre substrates or patterned resorbable polymers, cell speed and persistence of motion can be linked with directional cues in order to optimize the cellular velocity.

Delivery of spatial or directional cues to cells in the form of chemical gradients through 3D constructs requires that the chemical agent not only stimulates the cell function, but also that it is presented to cells as a directional gradient. A number of growth factors are chemotactic and certain extracellular macromolecules, such as fibronectin and collagen, produce chemotactic fragments on breakdown. Such diffusible agents can be important in tissue organization. Angiogenesis, for example, is based on the concept that new vessels move along gradients of diffusible factors. These can be towards sites of low oxygen tension or high lactate concentration, or towards local sources of growth factor production. For example, *vascular endothelial cell growth factor* (VEGF), or FGF, or matrix components such as oligosaccharide fragments of hyaluronan influence angiogenesis.

Whilst such gradients may be useful in special circumstances, their use in control of architecture is likely to be limited *in vitro*. The most plausible circumstances would be for:

1. Control over short ranges (e.g. between distinct cell layers in a tissue construct).
2. During early stages of matrix development and maturation.

Use of hydrated 3D matrices for support of cultured cells, based on collagen, fibrin, or fibronectin is helpful in minimizing convection mixing of gradients. However, as discussed earlier there are likely to be major practical difficulties in producing and maintaining effective gradients in a controlled manner as the matrix becomes dynamic, dense and anisotropic.

Mechanical Cues

A wide range of cell types are sensitive to mechanical loading. The mechanisms by which cells respond to mechanical signals are complex and include modulation of cytosolic-free calcium, stretch-sensitive ion channels, and deformation of cytoskeletal or integrin components. Responses can include changes in cell alignment, synthesis of active regulatory molecules such as growth factors or hormones, altered matrix synthesis, and enzyme release.

There are at least three different forms of mechanical cue: tension, compression, and shear. Fibroblasts are normally considered in terms of tensional forces, vascular endothelial cells predominantly under fluid shear, and chondrocytes are adapted to compression forces. It is possible to further differentiate each of these forms of mechanical cue, for example into cyclical and static force or on the basis of loading velocity. However, our understanding of the effects of these forms of loading or their combinations is limited.

A further, and potentially critical complexity, is that mechanical stimuli used in *in vitro* experimental models frequently have little or no directional component. Systems which give either non-orientated, or highly complex, multiple direction cues can give little information on the control of architecture and tissue polarity. They are valuable primarily for investigation of the role of mechanical forces on tissue production, turnover, and composition.

A number of shear force models are being developed for tissue engineering of tube tissues such as blood vessels. Fluid shear forces have been found to operate through cytoskeletal changes, cell-cell adhesion sites, and protein kinase and G protein signalling in vascular endothelial cells. These pathways are reported to affect heat shock protein phosphorylation, Cu/Zn superoxide dismutase activity, and apoptosis. Alteration of extracellular matrix productionand cell shape and alignment are more likely to be important for microarchitecture control cues. Current examples of tissue engineered tube structures for use in fluid drainage are bioartificial blood vessels and bladder.

The ability of fibroblasts to become orientated along a uniaxial mechanical load affects their pattern of matrix remodelling. Similar orientations have been reported in loaded skeletal muscle cells. Multi-axial complex loading patterns, produced by the flexcell analytical model, produces complex cell alignments, which are difficult to interpret, though collagen production is quantitatively increased by loading. In uniaxial, cyclically loaded collagen lattices, fibroblasts take on an alignment parallel with the maximum strain induced in the material at any given location. This has been used in one experimental model to identify two distinct zones of mechanical loading in the same gel.

Most studies of compressional loading have used chondrocytes in a 3D agarose gel matrix or whole cartilage. The agarose matrix provides little or no attachment until chondrocytes have elaborated their own pericellular matrix. In addition, agarose is essentially homogeneous in structure, providing uniform mechanical support in all dimensions. Consequently, newly synthesized matrix is easily detected and applied loads act uniformly and predictably on almost all the cells. In contrast, cartilage in organ culture contains cells in a non-homogeneous matrix with mechanically distinct zones.

Protocols

Cell Seeding of Implantable Materials

There are two main types of cell seeding strategy, depending on the final function of the implant:

1. Materials that have cells throughout their structure.
2. Materials that have cells on one or more surfaces.

There will also be situations which call for combinations of both 1 and 2. The method of cell seeding throughout an implant will depend on the properties of the substrate used. These fall into two basic categories:

1. 3D substrates which are preformed, e.g. polymers (PGA and PIA), collagen sponges, fibronectin mats.
2. Substrates which require some form of preparation, such as mixing or gelling, e.g. fibrin sealant, collagen gel.

Equipment and reagents

1. Sterile scalpels and forceps, hDF cells: 10^4 cells suspended in 0.5 ml.
2. DPPA non-crosslinked collagen sponges DMEM/1 cm^2 surface area.
3. DMEM supplemented with 10% fetal calf serum, 2 mM L-glutamine

Method

1. Cut sponge to desired size, shape under sterile conditions, and place each piece in a separate well in the multiwell plate.
2. Slowly add the medium to hydrate the sponge.

3. When the sponge is completely saturated (approx. 2 h), gently squeeze out the medium with a pair of flat forceps without disturbing the architecture of the sponge.
4. Add cell suspension dropwise onto the sponge, which will soak up the cell suspension and disperse the cells within the sponge.

Slow Release Systems for Local Control of TE Constructs or Repair Sites

Storage and release of diffusable agents from tissue engineering materials can be useful. Phenytoin (5,5-diphenylhydantoin, PHT) is an anti-convulsive agent which influences tissue matrix deposition. The drug is deposited into a fibronectin mat and, since it has very poor water solubility, it is trapped in crystal form.

Phenytoin Release from Fibronectin Mats

Equipment and reagents

1. Cut materials (e.g. fibronectin mats) into small squares of at least 5 mg mass.
2. 20 mg/ml PHT dissolved in 40% ethanol in distilled water. Phenol red-free DMEM, to monitor the release of PHT from the material

Method

1. Place dried, weighed pieces of material on the bottom of suitable cell culture wells.

The dynamics of PHT release tan be measured by covering the material with phenol red-free DMEM (200-500 μl). After various times, the medium is recovered and analysed for PHT by a routine quantitative HPLC assay. PHT release profiles from fibronectin mats.

In an alternative protocol, PHT can be delivered from a fibrin depot using a similar procedure, except that solid PHT is mixed with the fibrinogen component of a commercial fibrin glue preparation (SNBTS) followed immediately by the normal volume of thrombin solution component, at which point a fibrin clot is formed, trapping the PHT.

Slow Release Growth Factor Depot Formation

Reagents

- Growth factors such as basic fibroblast growth factor (unmodified or ^{125}I-labelled, to measure rate of release; Amersham Radiochemicats), nerve growth factor, transforming growth factor β, neurotrophin-3, dissolved to appropriate concentrations in DMEM or PBS DMEM supplemented with 2% fetal calf serum, penicillin/streptomycin (50 U/ml and 50 μg/ml respectively) PBS: 10 mM sodium phosphate buffer pH 7.4, containing 0.15 M NaCl.

Method

1. Cut suitably sized, freeze-dried fibronectin mats, add the growth factor in solution for 24 h at 37°C; 500 ng/ml NT-3, 3-5 μg/ml NGF.
2. Rinse with fresh DMEM to remove unattached growth factor.
3. For release monitoring, sample the supernatant at 24 h intervals.

Soluble growth factors can be incorporated into materials by impregnation into dry materials. This technique relies on the rapid ingress of water, carrying the soluble factors, into the dry porous material, followed by slow diffusional release of the solutes into the surrounding aqueous medium or implant site. Governing factors include the dimensions of the material (i.e. diffusion path length), its mean pore diameter, and the molecular weight of the growth factor.

Release can be monitored by dorsal root ganglion outgrowth bioassay for the neurotrophic factors or release of radiolabelled growth factor. In most cases, there is little further release of factors after three to five days incubation at 37°C.

4

Cells in Artificial Environment

Kinetics of cell growth and metabolism is desired for a better understanding of cell physiology and for the optimization and control of animal cell cultures. Consequently, it has been the objective of a large number of investigations in the past. A large number of factors were reported to affect the kinetics of growth, death, and product formation of animal cells in *bioreactors*. Despite these efforts, uncertainties and even controversies often exist in the literature regarding the relative importance of these factors in different cell lines. This may be mainly due to the lack of quantitative knowledge of the significance of these factors and particularly of the range of concentrations or values in which these factors are important. A mathematical description of the interactions of different factors in kinetic models can be very useful to solve this problem because they make it possible to separate and identify the effects of different factors. It can help better in interpreting and understanding experimental results obtained under different conditions. Models can also be used to check different hypotheses regarding regulation mechanisms. In addition, they can be useful as a guide tool for developing model-based process control and monitoring strategies.

For these reasons, considerable efforts have been made in the past to develop mathematical models to describe the growth, death, and metabolism of animal cell culture. Some of the kinetic models were reviewed or compared by Tziampazis and Sambanis, Porther and Schafer, and Goergen et al. Whereas structured models may be attractive due to their general applicability, they suffer from the shortcomings of unknown or complicated mechanisms of regulation of cell growth and death. In addition, they contain a large number of model parameters. The identification and estimation of these parameters are often difficult. Therefore, most of the present models for animal cell culture are unstructured models.

This chapter first gives an introduction to the kinetic characterization of cell culture in batch and continuous operations. This is followed by a brief description of major factors affecting the kinetics of cell culture and their quantitative analysis. The last part of this chapter describes typical models for kinetic analysis and simulation. Future research needs in cell culture kinetics are then discussed. No effort is made in this chapter to cover the *kinetics* and *modeling* of adherent cells and insect cells. This chapter is also not intended to be a complete survey of kinetics of animal cell culture. Rather, it emphasizes the quantification of cell growth and metabolism and the derivation of rate equations and mathematical models for kinetic analysis and simulations.

KINETIC CHARACTERIZATINO OF CELL CULTURE

Kinetics of Cells in Batch Culture

Kinetics of animal cell culture has been extensively studied in batch culture. Advances of batch culture include simple operation and process monitoring, short cultivation time, and relatively stable kinetic characteristics of cell lines if recombinant cells are used.

Time profile of the concentration of cells, nutrients, and metabolites

Both the viable cell number and the specific growth rate reach a maximum during the cultivation and decline afterwards in all cultures. This decline in growth is not due to the exhaustion of any macronutrient components (e.g., *glucose* and *amino acids*) or the inhibition of metabolic byproducts (e.g., *lactate* and *ammonium*) because the decline of μ inmost cultures occurs in the early stage of cultivation. Normally, lactate concentration increases until glucose is depleted and decreases thereafter. The reconsumption of lactate is particularly obvious in cultures with low initial glucose concentration. Ammonium formation is closely correlated with the consumption of glutamine.

The initial concentrations of glucose and glutamine can significantly affect the time profile of batch culture. However, for most cell culture media with initial glucose and glutamine concentrations above a certain level (about 1 mM in the case of BHK cells), these macronutrients have no significant effect on the maximum viable cell concentration and the specific cell growth rate.

For cell growth in batch culture, the inoculum age and size can have significant influences on the *kinetic behavior*. Martial et al. showed that increasing the age of inoculum resulted in a longer lag phase, a reduced maximal cell concentration, and a lower maximum growth rate of hybridoma cells seeded with the same initial cell density. For a number of cell lines it has been demonstrated that the initial cell density significantly affects growth in batch culture and a critical inoculum size exists. Relatively less is known about the effects of cell density on the metabolism and product formation of batch culture. Ozturk and Palsson found no significant effects of initial cell density (10^2 to 10^5 cells/mL) on the nutrient uptake rate and *monoclonal antibody* (MAb) formation rate of a hybridoma batch culture. This seems to disagree with some of the results from continuous cultures.

Intracellular content and excreted protein component

The intracellular content of many intermediary metabolites, proteins, and nucleic acid may vary during batch culture and is particularly useful for kinetic characterization of growth and *metabolism*. The intracellular content may be studied either by cell disruption, extraction, and subsequent analysis of the different substances, or by in situ flow cytometry or NMR analysis. For example, Ryll and Wagner studied the variation of intracellular ribonucleotide pools during "*perfused*" batch cultivations of hybridoma and BHK cell lines. These authors found that three cell specific regularities, i.e., the NTP ratio (NTP=(ATP+GTP/(UTP+CTP)), the uridine (U) ratio (U=UTP/(UTP-GNAc)), and the combined ratio NTP/U, are particularly useful for characterizing the cell growth. For a hybridoma cell line, the NTP value reduces rapidly during the adaptation (lag) phase and remains at a nearly constant and low value during the exponential growth phase. During the phase of reduced growth rate it dramatically increases.The U-value changes in an opposite direction to the NTP value. The intracellular nucleotide pools also sensitively respond to the presence of growth inhibitor such as ammonium. Studies of different cell lines in batch, chemostat, and perfusion cultures suggest that the content of nucleotide pools is cell line specific. For a given cell line the intracellular nucleotide ratios can be used as a sensitive parameter for process monitoring and control.

The variation of the U-value is primarily due to the change of the content of UDP-N-acetylhexosamines (UDPGNAc pool, primarily

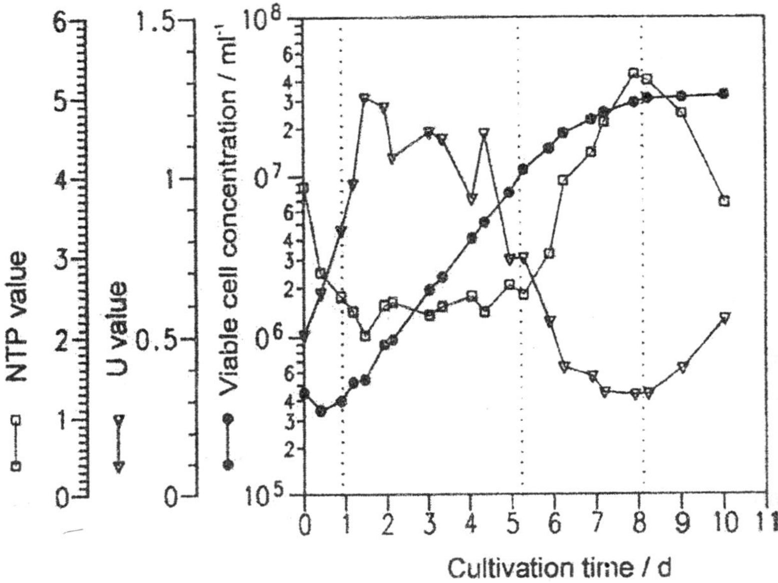

Fig. 4.1. Change of NTP and U values during the growth cycle of a "perfused" batch culture of a hybridoma cell line.

UDPG1cNAc and UDPGalNAc). Rapidly growing cells have a low UDPGNAc pool that rapidly increases when growth slows down. Since UDPGlcNAc and UDPGalNAc, as activated sugars, are precursors for the synthesis of oligosaccharides and present in the O-and N-glycanes of glycoproteins, they can influence the expression of carbohydrates in glycoproteins. Since the UDPGNAc pool significantly changes in batch culture, the glycosylation of excreted proteins can also vary during a batch culture. In particular, ammonium ion concentration can strongly affect the UDPGNAc pool and hence the glycosylation of excreted proteins. For a recombinant *Chinese hamster ovary* (CHO) cell line producing an immunoadhesin tumor necrosis factor-immunoglobulin it was shown that as ammonium increased from 1 to 15 mM, a concomitant decrease of up to 40% in terminal galactosylation and sialylation of the product occurred. The proportion of the nonglycosylated form of excreted protein seems to generally increase during batch culture. In addition to the changes of concentration of nutrients and metabolites, variations of glycosylation of secreted proteins can also be caused by modifications in the intracellular protein maturation processes or to the extracellular action of proteases or glycosidases.

Variation of cell morphology

Cell morphology (cell size and cell surface structure) significantly changes during batch culture. The cell size usually increases during the exponential growth phase but decreases during the stationary or decline phase. In the *late phase* of a batch culture, the cell surface (membrane) undergoes significant structural changes resulting in the disappearance of microvilli and the appearance of blebs. The cell size change is related to cell cycle and population change. The variation of cell morphology is correlated with the growth and death of cells. The increase of cell size in the growth phase is due to active cell proliferation with a high portion of cells in the S-phase of the cell cycle. On the other hand, the decline of cell size and changes of cell membrane toward the end of a batch culture can be explained by the mechanisms of cell death.

Mammalian cells principally die by two mechanisms: *necrosis* and *apoptosis* (also called *programmed cell death*). These two types of cell death display distinctive differences in their morphological and biochemical features. Morphologically, necrotic cell death is characterized by a progressive hydration of the cytoplasma (swelling) that is followed by membrane and organelle disruption, leakage of lysosomes into cytoplasm, nuclear disintegration, and finally complete

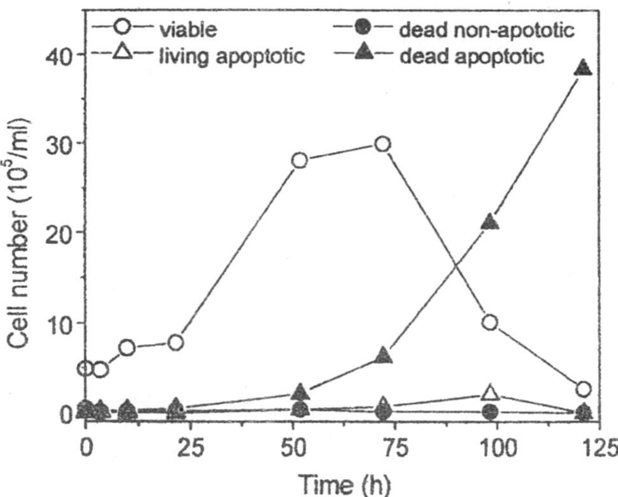

Fig. 4.2. Typical time profile of cell populations in a batch culture of the hybridoma cell line HyGPD YK-1-1 (N_t = total number, N_v = variable number, N_d = dead cell number, Vit = vitality) as determined by acridine orange/ethidium bromide staining.

disruption of the cell. The morphology of apoptotic cell death markedly contrasts with that of necrosis. Important features of apoptosis include condensation and fragmentation of chromatin, loss of volume (*cell shrinkage*), and modification of the *cytoskeleton*. Following these events the plasma membrane convolutes and the cell eventually fragments to form membrane-bound vesicles (*apoptotic bodies*) that contain intact cytoplasmic organelles and, usually, nuclear fragments. In the body or under *in vivo* conditions in tissues apoptotic bodies are recognized and engulfed by *phagocytic* or nearby cells. The recognition of apoptotic bodies has been suggested to be due to changes of the membrane including occurrence of the signal molecule phosphatidylserine on the membrane surface. A number of investigations have shown that in the physical and chemical environments of a bioreactor most production cell lines including hybridoma, myeloma, CHO, and *baby hamster kidney* (BHK) cells primarily die by apoptosis.

Determination of cell population heterogeneity for kinetic studies

For the study of cell culture kinetics it is important to quantify the different types of cells and to follow their variations during cultivation. The determination of cell population heterogeneity can be conventionally based on either morphological or biochemical features. For example, based on the properties of cell membrane and the structural morphology of cell nuclei the method of acridine orange and ethidium bromide staining can be used to distinguish four types of cells, namely *viable nonapoptotic* (VNA), *viable apoptotic* (VA), *nonviable apoptotic* (NVA), and *nonviable nonapoptotic cells* (NVNA) (i.e., *necrotic dead cells*). The relative numbers of both living and dead apoptotic cells are low at the beginning of the batch culture. Whereas the number of living apoptotic cells remains at a low level (maximum 2–5% of the total cells) throughout the cultivation, the number of apoptotic death cells increases continuously and accounts for more than 97% of the total dead cells at the end of the cultivation. These results demonstrate that, first, apoptosis is the prevailing death mechanism of this cell line and, second, the kinetics of transition of living apoptotic to dead cells is much faster than the kinetics of apoptosis induction.

Flow cytometry is also a very useful tool for studying cell population heterogeneity. By staining cells with proper fluorophores, such as propidium iodide and acridine orange, flow cytometry can be used to quantify the progress of cells through the cell cycle. During batch culture, the proportion of cells in the G1, S, and G2/M phases significantly changes. In the exponential growth phase, most of the cells are in the S and G2 phases, whereas during the stationary and decline phases the majority of cells are in the G1-phase. Flow cytometry can be also used to determine the populations of living, apoptotic, and dead cells by combination with propidium iodide and annexin staining.

In contrast to the large body of knowledge concerning the molecular mechanisms, biochemical features, and physiological consequences of apoptosis, relatively little is known about the *kinetics* and *dynamics* of cell death (for both *apoptosis* and *necrosis*). This may be mainly due to two reasons. First, apoptotic cell death is a relatively fast process. In vivo, it may take only a few hours from the early stage of induction (evanesce to recognition) to the end stages of *degradation* and *phagocytosis*. Second, reliable methods for quantification of apoptosis have only recently become available, particularly for suspension culture. Therefore, most of the studies dealing with cell death have been descriptive or qualitative. To better understand and intervene in the apoptotic process a quantitative and systematic knowledge of cell death is important. Linz reported on the kinetics of apoptosis in three mammalian cell lines (hybridoma, BHK, and CHO) and in the human carcinoma cell line HeLa in normal batch cultures and cultures under different stress conditions. Similarly, the hybridoma cell culture, apoptosis is the prevailing death mechanism of all these cell lines and the kinetics of transition of living apoptotic to dead cells is much faster than the kinetics of apoptosis induction.

Estimation of specific rates of cell growth, death, and metabolism

The specific mitotic rate as defined in Eq. (1) is normally used to calculate the specific rate of cell growth m (hr^{-1} or day^{-1}) in batch culture:

$$\mu = \frac{dN_t}{dt} \cdot \frac{1}{N_v} \qquad \ldots(1)$$

Similarly, Eq. (2) is often used to calculate the specific death rate (k_d) of cells:

$$k_d = \frac{dN_d}{dt} \cdot \frac{1}{N_v} \qquad \ldots(2)$$

In Eqs. (1) and (2), $N_{t,}$ N_v, and N_d are the total cell number, viable cell number, and dead cell number per unit culture volume, respectively. N_t, N_v, and N_d can be determined by counting the viable and dead cells using the trypan blue exclusion method with a hemocytometer. N_t can be also determined by counting the cell nuclei using the so-called CASY Counter after disruption of the cells. This method is particularly suitable for cultures with aggregate formation.

To better quantify the kinetics of apoptotic and necrosis cell death processes, Linz et al. introduced the terms apoptosis induction rate ($k_{\text{ap}}^{\text{in}}$) apoptotic death rate ($k_{\text{ap}}^{\text{death}}$) and necrotic death rate ($k_{\text{nec}}^{\text{death}}$):

$$k_{\text{ap}}^{\text{in}} = \left(\frac{dN_{NVA}}{dt} + \frac{dN_{VA}}{dt} \right) \times \frac{1}{N_{VNA}} \quad \ldots(3)$$

$$k_{\text{ap}}^{\text{death}} = \left(\frac{dN_{NVA}}{dt} \right) \times \frac{1}{N_{VA}} \quad \ldots(4)$$

$$k_{\text{nec}}^{\text{death}} = \left(\frac{dN_{NVNA}}{dt} \right) \times \frac{1}{N_{VNA}} \quad \ldots(5)$$

where N_{VNA}, N_{VA}, N_{NVA}, and N_{NVNA} are the number of viable nonapoptotic, viable apoptotic, nonviable apoptotic, and nonviable nonapoptotic cells per unit volume of culture, respectively.

The specific rates of consumption of major nutrients (e.g., glucose and glutamine) and formation of metabolites (e.g., lactate and ammonium) in batch culture can be calculated as:

$$q_{\text{Glc}} = -\frac{dC_{\text{Glc}}}{dt} \frac{1}{N_v} \quad \ldots(6)$$

$$q_{\text{Lac}} = \frac{dC_{\text{Lac}}}{dt} \frac{1}{N_v} \quad \ldots(7)$$

$$q_{\text{Gln}} = \left[-\frac{dC_{\text{Gln}}}{dt} - kC_{\text{Gln}} \right] \frac{1}{N_v} \quad \ldots(8)$$

$$q_{\text{Amm}} = \left[\frac{dC_{\text{Amm}}}{dt} - kC_{\text{Gln}} \right] \frac{1}{N_v} \quad \ldots(9)$$

In Eqs. (8) and (9), k is a rate constant accounting for the spontaneous decomposition of glutamine as a first order reaction and having a value of 4.675×10^{-3} hr^{-1} at 37°C.

To calculate the specific rates it is recommended to first fit the experimental data to appropriate polynomial functions by the least-squares method. For these purposes, experimental data with relatively short time intervals are required. The time course of some variables sometimes has to be divided into several sections to find functions describing the data satisfactorily. The derivatives with respect to time of these functions are then used to calculate the specific rates of cell growth, death, and metabolism.

The time courses of μ,(k_{ap}^{in}), (k_{ap}^{death}), and (k_{nec}^{death}) for the batch culture shown. It is interesting to note that whereas (k_{ap}^{death}) is more or less constant throughout the cultivation, (k_{ap}^{in}) is quite low during the growth phase and significantly increases during the death phase. In general, (k_{ap}^{death}) is much higher than (k_{ap}^{in}) (particularly during the growth phase) and m. The fast kinetics of apoptotic cell death in comparison to the induction explains why only a very low population of living apoptotic cells can be experimentally determined. Because of this fast kinetics of apoptotic cell death and because cells of some cultures undergo a fast secondary necrosis after apoptotic death, these cultures may not display the so-called typical hallmarks of apoptotic cell death such as DNA fragmentation and hence are sometimes considered not to die by apoptosis. A careful and fast measurement of many parameters is important for a reliable kinetic characterization of cell death.

The specific rates of metabolism of the batch culture follow a similar trend as the specific cell growth rate. This gives the impression that cell growth was limited by the availability of macronutrients. However, the decline of specific cell growth rate cannot be overcome by simply increasing the nutrient concentration, indicating that other factors are involved. The relationships between the specific rates of growth and metabolism will be discussed in the following section.

Results from kinetic study of batch cultures suggest that the specific consumption rate of glucose and specific formation rate of lactate are closely related. This also seems to apply to the relationships between the specific consumption rate of glutamine and the specific formation rate of ammonium. Zeng et al. analyzed the stoichiometric ratios of a number of cell lines under diverse experimental conditions including batch cultures.

Cell Kinetics in Continuous Culture

Advantages and application of continuous culture

The advantages and potentials of continuous culture for cell culture technology can be viewed from two general aspects. From an application point of view, these include high volumetric productivity, savings in labor and energy costs (e.g., inoculum preparation, reactor cleaning, and sterilization), uniform product quality, better automation and process control, and the use of more efficient and economic methods of medium preparation and downstream processing. If cell recycling (perfusion culture) is applied, the concentrations of cells and product can be significantly improved. As a research tool, continuous culture provides well-defined cultivation conditions for genetic, biochemical and physiological characterizations of cells. It allows an independent variation of growth parameters, enabling reliable kinetic studies of cell growth and metabolism for process optimization. The transition behavior of a continuous culture upon shift-up or shift-down of a variable is also a powerful tool to study the regulation of growth and metabolism. In addition, continuous culture is a useful means for selecting strains and=or subclones with improved growth characteristics and productivity. On the other hand, the mutation and instability of some producing strains due to the selection pressure in continuous culture is a major obstacle for the industrial application of this cultivation technique, particularly in the case of genetically modified cells.

Continuous operation is used for the production of monoclonal antibodies and recombinant proteins by animal cell culture. The major applications of continuous culture are however still found in fundamental studies and process optimization in laboratory scale. Animal cell cultures have been quantitatively studied in depth with the help of continuous culture.

Time profile and steady state of continuous culture

A typical time profile of total cell number (N_t) determined by counting cell nuclei, the cell viability, and the concentrations of glucose, lactate, glutamine, and ammonium of two series of continuous cultures at a constant dilution rate but with varied concentrations of glucose and glutamine in medium.

Steady states with varied residual concentrations of nutrients and hence varied growth limitation were approached at relatively constant specific growth rates. The steady states had altered consumption rates of nutrients and formation rates of products. In the experiment series C1, N_t increased up to 5 × 10^6 cells/mL with increasing glutamine in the feed ($C_{\mathrm{Gln}}^{\mathrm{in}}$) at low concentration range but leveled off above a glutamine concentration of 3.5 mM. The residual glucose concentration (C_{Glc}) was very low (between 0.20 and 0.27 mM) throughout the cultivation. The residual glutamine concentration (C_{Gln}) was also very low (between 0.04 and 0.16 mM) at low $C_{\mathrm{Gln}}^{\mathrm{in}}$ (between 1 and 3.5 mM). Under these conditions, growth may be partly limited by glucose and/or glutamine. A further increase of $C_{\mathrm{Gln}}^{\mathrm{in}}$ from 3.5 to 5 mM led to an increase of C_{Gln} from 0.14 to 0.31 mM. Since there was no increase of cell number in this range of $C_{\mathrm{Gln}}^{\mathrm{in}}$ it can be assumed that growth was not limited by the availability of glutamine under these conditions. The increased consumption of glutamine was accompanied by a corresponding increase of ammonium production. Lactate concentration was below 1 mM throughout the cultivation. In the culture series C2, N_t increased with increasing glucose in the feed ($C_{\text{Gln}}^{\text{in}}$) at low concentration range as well. It reached a maximum value of 11×10^6 cells/mL at an initial glucose concentration of 6 mM and decreased afterwards with further increases of $C_{\text{Gln}}^{\text{in}}$. The residual glucose concentration was in a similar range to that of culture C1 (0.18–0.29 mM) at $C_{\text{Gln}}^{\text{in}}$ below 10mM. At $C_{\text{Gln}}^{\text{in}}$ = 10 mM, C_{Glc} increased to 1.36 mM and was obviously not growth limiting. Throughout the cultivation, C_{Gln} was in the limitation range (between 0.04 and 0.13 mM) as found in culture series C1. Culture series C2 was accompanied with an increased production of lactate. In contrast, the ammonium concentration decreased.

Estimation of specific rates of cell metabolism in continuous cultures

Steady-state data from continuous culture can be used to calculate the specific rates of cell metabolism under clearly defined physiological conditions. Under steady-state conditions, the specific rates of cell growth, death, nutrient consumption, and product formation of a conventional continuous culture can be calculated as follows:

$$\mu = D\frac{N_t}{N_v} \quad ...(10)$$

$$k_d = \mu - D \quad ...(11)$$

$$q_{\text{Glc}} = (C_{\text{Glc}}^{\text{in}} - C_{\text{Glc}})\frac{D}{N_v} \quad ...(12)$$

$$q_{\text{Lac}} = (C_{\text{Lac}} - C_{\text{Lac}}^{\text{in}})\frac{D}{N_v} \quad ...(13)$$

$$q_{\text{Gln}} = [(C_{\text{Gln}}^{\text{in}} - C_{\text{Gln}}) \cdot D - k \cdot C_{\text{Gln}}]\frac{1}{N_v} \quad ...(14)$$

$$q_{\text{Amm}} = [(C_{\text{Amm}} - C_{\text{Amm}}^{\text{in}}) \cdot D - k \cdot C_{\text{Gln}}]\frac{1}{N_v} \quad ...(15)$$

where D is the dilution rate (hr^{-1} or day^{-1}); C_j^{in} and C_j are the concentrations of substance *j* in the feed and in the culture, respectively.

In a perfusion culture (cell recycle) with complete cell separation (i.e., by means of membrane filtration) a cell-free permeation stream is withdrawn at a flow rate of F_p. A stream of culture with cells at the same concentration as in the reactor is also withdrawn at a flow rate of F_B. The perfusion rate (*P*) of the culture is an important process variable:

$$P = \frac{F_P + F_B}{V} \quad \text{...(16)}$$

where V is the reactor volume of the culture. In this case, the cell bleed rate (D_B) is used for the calculation of cell growth and death rates:

$$D_B = \frac{F_B}{V} \quad \text{...(17)}$$

$$\mu = D_B \frac{N_t}{N_v} \quad \text{...(18)}$$

$$k_d = \mu - D_B \quad \text{...(19)}$$

Thus, the perfusion rate *P* should be used instead of the dilution rate *D* in Eqs. (12)-(15). In practice, cell retention devices such as centrifuges and spin filters may not completely separate cells from culture broth. In such a case, the above equations need further modifications. The term retention ratio (*r*) may be used to describe the ratio of cell number in the permeate to the total cell number in culture broth:

$$r = \frac{N_P}{N_t} \quad \text{...(20)}$$

where N_p is the cell number per unit permeate volume. A modified form of Eq. (17) is needed to describe the real or effective cell bleed rate (D^*_B):

$$D_B^* = \frac{F_B + F_P \cdot r}{V} = D_B(1-r) + rP \quad \text{...(21)}$$

For high cell density perfusion culture cell lysis may considerably contribute to the death of cells. In such a case, caution should be taken in estimating the total cell number N_t in the above equations. The measurement of release in the culture medium of the cytoplasmic enzyme lactate dehydrogenase has been suggested as a means to estimate the total cell number.

Influence of Envirinmentla and Physiological Conditions and Rate Equations

A large number of factors have been reported in the literature to affect the kinetics of growth, death, metabolism, and product formation of animal cells in bioreactors.These include environmental physicochemical parameters, composition and levels of nutrients, growth factors and the formation of metabolites, and some not yet identi fied autocrine factors. In this section, only some of these factors are addressed in view of kinetic analysis and optimization of cell culture. More general information about the effects of some of the factors and cell physiology can be found in this book.

Effect of Nutrients and their Consumption Rates

Glucose and glutamine are the two major nutrients for animal cell culture that can strongly affect the kinetics. Their effects on the kinetics of cell culture have been shown for the batch and continuous cultures of BHK cells, respectively. The steady-state specific rates of cell growth, nutrient consumption, and product formation for the continuous culture under varied initial glucose and glutamine concentrations in medium. The specific growth rates of the continuous cultures are relatively constant (1.0–1.2 day^{-1}).

Despite a relatively constant growth rate, the cultures exhibit strongly varied specific rates of nutrient consumption and metabolite formation. As found for several other cell lines in batch and continuous cultures, q_{Glc} strongly depends on the residual glucose concentration. For the BHK cell line it begins to significantly increase at $C_{Glc} > 0.2$ mM. Similarly, q_{Gln} is mainly affected by the residual glutamine concentration and begins to increase notably at $C_{Gln} > 0.1$ mM except for the culture C2-4, which has an obvious glucose excess. q_{Lac} is somewhat less sensitive to glucose compared to q_{Glc}. The former begins to increase significantly at $C_{Glc} > 1$ mM. q_{Amm} is similarly sensitive to glutamine as q_{Gln}.

Obviously, the linear maintenance model previously used to describe the nutrient consumption of animal cells cannot be generally applied. Based on the observation that the nutrient consumption rate is not only dependent on cell growth rate but also strongly on the nutrient concentration, Zeng and Deckwer proposed the following kinetic expressions for the rates of consumption of glucose and glutamine of animal cells:

$$q_{Glc} = m_{Glc} + \frac{\mu}{Y_{Glc}^{max}} + \Delta q_{Glc}^{Glc(max)} \frac{C_{Glc} - C_{Glc}^{*}}{C_{Glc} - C_{Glc}^{*} + K_{Glc}^{Glc}} + \Delta q_{Gk}^{Gln(max)} \frac{C_{Gln} - C_{Gln}^{*}}{C_{Gln} - C_{Gln}^{*} + K_{Glc}^{Gln}} \quad \text{...(22)}$$

$$q_{\rm Gln} = m_{\rm Gln} + \frac{\mu}{Y_{\rm Gln}^{\rm max}} + \Delta q_{\rm Gln}^{\rm Glc(max)} \frac{C_{\rm Gln} - C_{\rm Gln}^*}{C_{\rm Gln} - C_{\rm Gln}^* + K_{\rm Gln}^{\rm Glc}}$$

$$+\Delta q_{\rm Gln}^{\rm Gln(max)} \frac{C_{\rm Gln} - C_{\rm Gln}^*}{C_{\rm Gln} - C_{\rm Gln}^* + K_{\rm Gln}^{\rm Gln}} \qquad \text{...(23)}$$

for $C_{\rm Glc} \geq C^*_{\rm Glc}$ and $C_{\rm Gln} \geq C^*_{\rm Gln}$, Eqs. (22) and (23) express the substrate consumption rates as a sum of the substrate consumption rate under substrate limitation and two additional consumption rates owing to the excess of glucose and glutamine, respectively. Similarly to the substrate consumption model of microbial cells under substrate-sufficient conditions, Eqs. (22) and (23) may be called "*excess kinetics*" of animal cells. $m_{\rm Glc}$ and $m_{\rm Gln}$ are the maintenance requirements for glucose and glutamine that are often negligible in conventional batch and continuous cultures. $C^*_{\rm Glc}$ and $C^*_{\rm Gln}$ are defined as the concentrations of glucose and glutamine under dual limitations. They are often very small compared to concentrations of glucose ($C_{\rm Glc}$) and glutamine ($C_{\rm Gln}$) in the culture and can be neglected. $Y_{\rm Glc}^{\rm max}$, $Y_{\rm Gln}^{\rm max}$, $\Delta q_{\rm Glc}^{\rm Glc(max)}$, $\Delta q_{\rm Glc}^{\rm Gln(max)}$, $\Delta q_{\rm Gln}^{\rm Gln(max)}$, $\Delta q_{\rm Gln}^{\rm Glc(max)}$, $K_{\rm Glc}^{\rm Glc}$, $K_{\rm Glc}^{\rm Gln}$, $K_{\rm Gln}^{\rm Gln}$, and $K_{\rm Gln}^{\rm Glc}$ are constants with different physiological meanings.

Equations (22) and (23) satisfactorily described experimental data from batch, fed-batch, and continuous cultures of several animal cell lines. For the continuous culture following rate equations were established:

$$q_{\rm Glc} = \frac{\mu}{3.19} + 0.25\frac{C_{\rm Glc} - 0.06}{C_{\rm Glc} + 7.74} - 0.017\frac{C_{\rm Gln} - 0.07}{C_{\rm Gln} + 0.59}$$

$$(\text{mmol} / 10^9 \text{ cells h}) \qquad \text{...(24)}$$

$$q_{\rm Gln} = \frac{\mu}{3.19} + 0.044\frac{C_{\rm Gln} - 0.05}{C_{\rm Gln} + 0.22}(\text{nmol} / 10^9 \text{ cells h}) \qquad \text{...(25)}$$

Using the rate equations described above the interaction and regulation of glucose and glutamine utilization of animal cells were quantitatively analyzed. The results indicated that, whereas $q_{\rm Glc}$ is affected by glutamine, $q_{\rm Gln}$ appears to be not or less significantly affected by glucose. This is in accordance with the flux and enzyme analyses of metabolism of different animal cell lines reported by Neermann and Wagner. The authors found that the flux from pyruvate to acetyl-CoA is absent or very weak due to the lack of the enzyme pyruvate dehydrogenase. The previously reported effect of glucose on glutaminolysis in literature seems to be attributed to the indirect influence of glucose on cell growth and residual glutamine concentration. On the other hand, it is shown that glutamine can have positive or negative influences on the glycolysis, depending on the nature of growth limitation. The kinetic analysis also indicated that the utilization rate of glucose and glutamine is mainly affected by the residual concentrations of the respective compounds and less by the growth rate. Similar kinetic expressions as Eqs. (22) and (23) may also be used for the uptake or consumption of other major nutrients.

Effect of Metabolic Byproducts and their Formation Rates

Ammonium and lactate are two major byproducts in mammalian cell cultures. Lactate excretion is due to incomplete oxidation of glucose in the glycolysis pathway; pyruvate (the endproduct of this pathway) is transformed into lactate to maintain the oxidative state of the cell. The main source of ammonium formation is amino acid metabolism, particularly glutamine which serves as a protein constituent and also as the main energy source. At low concentration of glutamine, the consumption of other amino acids, especially essential amino acids, can signiflcantly contribute to the formation of ammonium.

Ammonium and lactate are often considered to be the dominating factors inhibiting cell growth and indeed have been shown to be toxic above certain concentrations. For example, ammonium level >2 mM or lactate level >20 mM were reported to inhibit cell growth and MAb production of hybridoma and CHO cells. The toxic action of lactate is probably due to the effects of pH and osmolarity of culture medium at relatively high concentration. An elevated ammonium concentration may reduce metabolic efflciency by forcing excretions of potentially valuable intermediate metabolites, such as alanine, to achieve ammonium detoxification. High concentrations of lactate and ammonium up to the critical levels can occur at the end of batch culture or in high cell density cultures. Therefore, removing or reducing lactate and ammonium formation has often been suggested as an important goal of process optimization. However, as shown by Zeng et al. and discussed in section "Rate laws of cell growth and death," the inhibitory effects of ammonium and lactate may be overestimated or even misinterpreted in many cases.

The formation rate of ammonium and lactate is mainly determined by the nutrient concentrations and growth rate. In animal cell culture there are clear stoichiometric relationships between the consumption of nutrients and formation of metabolites. Rate equations for the formation of the two major metabolites lactate and ammonium were therefore proposed which have a similar structure like Eqs. (22) and (23):

$$q_{\text{Lac}} = m_{\text{Lac}} + \frac{\mu}{Y^{\text{m}}_{\text{Lac}}} + \Delta q^{\text{m}}_{\text{Lac}} \frac{C_{\text{Gln}}}{C_{\text{Gln}} + K^{\text{Glc}}_{\text{Lac}}} \quad \text{...(26)}$$

$$q_{\text{NH}_3} = m_{\text{NH}_3} + \frac{\mu}{Y^{\text{m}}_{\text{NH}_3}} + \Delta q^{\text{m}}_{\text{NH}_3} \frac{C_{\text{Gln}}}{C_{\text{Gln}} + K^{\text{Gln}}_{\text{NH}_3}} \quad \text{...(27)}$$

The parameters m_{Lac}, m_{NH_3}, $Y^{\text{m}}_{\text{Lac}}$; $Y^{\text{m}}_{\text{NH}_3}$; $\Delta q^{\text{m}}_{\text{Lac}}$; $\Delta q^{\text{m}}_{\text{NH}_3}$; $K^{\text{Glc}}_{\text{Lac}}$ and $K^{\text{Gln}}_{\text{NH}_3}$ are constants having similar physiological meanings as those of Eqs. (22) and (23).

From kinetic analysis of nutrient consumption and byproduct formation it can be concluded that for an effective utilization of nutrients and for a reduction of toxic byproducts the glucose and glutamine levels should be controlled at low levels (e.g., ca. 0.1–0.5mM depending on cell lines). The critical nutrient levels may also depend on cell density under certain conditions. For example, Ljunggren and Haggstrom investigated glucose and glutamine-limited fed-batch cultures of hybridoma cells and found that glucose and glutamine concentrations below 1mM did not limit cell growth. Glucose limitation alone did not reduce ammonium formation, in comparison to a reduced ammonium release by about 50 and 80% caused by glutamine limitation and dual glucose and glutamine limitation, respectively. The metabolism of glucose and amino acids of a recombinant BHK cell line at low levels of nutrient concentration (glucose concentration <5 mM and glutamine concentration <1 mM) was quantitatively characterized by Linz et al. It was demonstrated that the uptake rates of glucose and glutamine are markedly reduced at low levels of glucose and glutamine, resulting in a more efficient energy metabolism and biosynthesis, and a reduced formation rate of lactate and ammonium.

Effect of Temperature

Cultivation temperature can significantly influence the growth and metabolism of animal cell cultures. Animal cell cultures are normally carried out at physiological temperature (37°C). Temperature stress (heat shock) was known to adversely affect many functions in animal cells. The most common cellular responses to heat shock include changes in gene expression, resulting in the synthesis of specific stress proteins, i.e., *heat-shock proteins* (HSPs). HSPs are a group of proteins that are highly conserved in all organisms from bacteria to mammals, many of which act as molecular chaperones for protein folding.

Recent studies focusing on the response of cells to decreased temperatures led to the discovery of cold inducible proteins. However, unlike HSPs, cold-shock proteins are not conserved among all species. Lowering the culture temperature generally suppresses cell growth but its effects on cellular productivity are variable among different cell lines and expression systems. It has been reported that a temperature shift from 37 to 30–33C can lead to a reduced growth rate, prolong the total generation time, and increase the viability of cultured mammalian cells. Cold-induced growth arrest was found to be G1-phase specific in CHO batch cultures. The prolonged culture viability at lower temperatures is considered to be a result of delayed onset of apoptosis that causes a rapid decrease in the percentage of cells in S-phase in CHO batch culture. CHO cells engineered to synthesize *secreted alkaline phosphatase* (SEAP) were also characterized by shifting the cultivation temperature from 37 to 30°C. This temperature shift resulted in a growth arrest mainly in the G1-phase of the cell cycle and a concomitant increase of specific productivity. However, the effects of low temperature on the protein production rate of mammalian cells seem to depend on cell lines and protein products. For example, Sureshkumar and Mutharasan reported that hybridoma cells cultivated at low temperatures had a reduced specific productivity of MAb. However, Kaufmann et al. found an enhanced SEAP productivity in CHO cells grown at 30°C. An up to 1.7-fold higher specific productivity and prolonged culture viability resulted in an overall 3.4-fold higher product yield in low-temperature cultivations compared to standard cultivations at 37°C. Although several reports suggest that controlled proliferation increases the productivity of mammalian cells, its influence on product quality should be investigated to better evaluate the potential of this strategy. The majority of pharmaceutically important proteins produced with mammalian cells require post-translational modifications for full therapeutic efficacy. Glycosylation is a particularly critical parameter for product quality because oligosaccharide structures can influence the solubility, stability, bioactivity, immunogenicity, and pharmacokinetics of a pharmaceutical protein. In this connection, it may be stated that cultivations at lower temperatures are favorable because of the more consistent quality (glycosylation and molecule fragmentation) and improved molecule integrity of protein products.

Culture temperature also affects the metabolic rates of animal cells. Low temperatures normally reduce cellular metabolic activities as reflected by the decreased specific rates of glucose uptake, lactate production, glutamine uptake, ammonium production, oxygen uptake, and CO_2 evolution. The decreased oxygen demand at lower temperatures makes it possible to support a higher cell concentration in a bioreactor. Furthermore, lower temperatures also result in a significant decrease in the $NaHCO_3$ addition for pH control.

Ludwig et al. studied the influence of temperature on the shear sensitivity of adherent BHK-21 cells grown at temperatures between 28 and 39°C. It was found that decreasing the temperature lowered the growth rate and increased the shear resistance in BHK cells. Cell morphology also changed at low temperatures. At 28°C the cells tented mostly to be more spherical or triangular, as opposed to a confluent monolayer at 37°C. This phenomenon was explained by an increased rigidity of the lipid bilayer of cell membrane.

Weidemann et al. compared batch and repeated-batch cultures of BHK-21 cells for the production of AT III at 33 and 37°C. At these temperatures, the speciflc growth rate of the cells was 0.50 and 0.62 $days^{-1}$, while the specific glucose uptake rate was 0.45 and 0.58 ng/cell/day, respectively. A higher product titer was reached at 33°C, a temperature at which the medium demand can be rationalized due to reduced nutrient consumption.

Jorjani and Ozturk quantitatively studied the effect of temperature on three different mammalian cell lines (BHK, murine hybridoma, and CHO). In all cases, the specific *oxygen uptake rate* (OUR) qO_2 decreased by about 10% for one degree of reduced temperature. The effect of temperature was observed to be exponential and can be well described by Arrhenius' equation:

$$qO_2 = qO_2^0 \cdot e^{-E/RT} \qquad \text{...(28)}$$

where E is the activation energy, R is the ideal gas constant, and T is the absolute temperature in degrees Kelvin. E was found to be similar for different cell lines (between 80 and 90 kJ=mol), indicating a similar mechanism for the effect of temperature on oxygen consumption. The effects of temperature on cell growth, nutrient consumption, and formation of metabolites and a protein product in high-density perfusion cultures were also quantitatively studied for the temperature range 34–37°C. Whereas the nutrient consumption and formation of lactate and ammonium decreased with decreasing temperature in a nearly linear manner, the formation of the protein product increased. Kinetic expressions for these observed effects have not been established yet. It may be possible that Arrhenius equation can also be applied to describe the temperature effects on primary cell metabolism.

In conclusion, decreasing the culture temperature from 37 to about 30–34.0°C appears to be advantageous for many animal cell cultures, especially for high-density perfusion cultures of animal cells.

Effect of pH

pH affects the metabolism, growth, and protein production of animal cell culture in various ways. The effects of pH on cell metabolism and growth have been examined in cell lines of HeLa, lymphoblastoid, hybridoma, and HL60. The optimum pH for growth varies with cell lines. It is worth mentioning that a pH excursion as small as 0.2 pH units can profoundly influence the growth, metabolism, and productivity of animal cells in some cases. Problems in product quality and uniformity may also arise if the culture pH is not properly controlled. In fact, an effective pH control is not always obtainable in cell culture, especially in large-scale bioreactors, since spatial heterogeneities in the culture system may exist due to imperfect mixing or cell density effect that results in a pH gradient. For example, Akatov et al. showed that in T-flask cells of Chinese hamster fibroblasts had a local pH of 6.5, although the pH of the bulk liquid remained at 7.6.

Osman et al. investigated the effect of pH shifts on cell growth and productivity in batch culture of a GS-NSO mouse myeloma cell line. They reported that pH shifts above 0.2 units caused a transient increase in apoptotic cell death. However, cultures shifted to pH values between 7.0 and 8.0 continued to grow and the apoptotic fraction returned to the initial levels. Cultures shifted to pH values above pH 8.0 and below pH 7.0 did not recover, resulting in cell death. After the pH shift, a maximum specific growth rate was observed over the range of pH 7.3–7.5 and the maximum viable cell number was observed at pH 7.3. A maximum volumetric antibody production, resulting from increased culture longevity, was found at pH 7.0. It was also observed that glucose consumption rate increased with increasing pH. Exposure of cells to a pH value >8.5 for more than 10 min caused a decrease in the proportion of viable cells and induced a lag phase in cell growth.

pH can influence the structure of proteins either directly or indirectly by influencing the cellular glycosylation pathways. For example, the extracellular pH value was found to affect the glycoform distribution of IgG from hybridomas and mPL-I in CHO cells. Glycoforms of mPL-I with similar molecular sizes were expressed between pH values of 7.2 and 8.0. However, a decreased glycosylation of mPL-I occurred at both lower pH (pH < 6.9) and higher pH (pH

> 8.22). The intracellular pH (pHi) plays a decisive role in mediating the pH effects mentioned earlier. It can be conveniently measured by using flow cytometry with carvoxy-SNARF-1. Ishaque and Al-Rubeai determined the relationships among pHi, apoptosis, and cell cycle in hybridoma cells. It is reported that temporal changes in the distribution of proliferative capacity (S-phase), metabolic activity (pHi), and cell death population dynamics can be effectively and reliably determined using this approach. It is shown, for example, that intracellular acidification precedes the occurrence of apoptosis during batch culture, suggesting that the decrease in pHi can be used as an indicator of cellular deterioration and cell death. pHi has also been reported to affect the localization of hexokinase, resulting in an increase in glycolysis with increasing pHi.

Dissolved Oxygen and Oxygen Uptake Rate

Dissolved oxygen concentration (DO or pO_2) is an important variable for mammalian cell culture. Oxygen is essential for the efficient generation of ATP and must be continually supplied to cultured cells due to the low solubility of oxygen in culture medium (7.8 mg/L in water at 25°C). Elevated oxygen concentration (usually above 100% of air saturation), or hyperoxia, has a negative effect on cell growth. Hyperoxia can damage cellular macromolecules such as DNA, proteins, and lipids. This can, in turn, damage the cell membrane and induce cell death. The toxicity of oxygen is associated with the production of intracellular *reactive oxygen species* (ROS) that damage a variety of cellular components or impair their functions.

Dissolved oxygen also has profound effects on cellular metabolism. A frequently observed trend is that once DO drops below a critical value (typically between 1 and 10% of air saturation) the oxygen consumption rate decreases and the specific glucose consumption and lactate production rates increase. The optimum DO value for antibody production differs from that for cell growth. Miller et al. reported an increase in the total and viable cell number when DO was decreased to a critical value of 0.5% air saturation with continuous culture of a hybridoma cell line, while the optimum DO for antibody production was 50% air saturation. Also Ozturk and Palsson investigated the effects of DO on hybridoma cell in a continuous stirred tank bioreactor where the DO value was varied between 0 and 100% of air saturation. Cell growth was inhibited at both high and ow DO values. Cell viability was higher at low DO. Interestingly, they showed that the consumption rates of glucose, glutamine, and oxygen remained relatively constant at DO values above 1% of air saturation. But when growth became oxygenlimited (DO 1–0%), the consumption rates of glucose and glutamine increased by 2–3-folds and a corresponding increase of formation of lactate and ammonium was observed. It was concluded in this study that the antibody concentration was the highest at DO=35% of air saturation, while the specific antibody production rate was insensitive to DO.

While a low level of oxygen concentration (e.g., below 10% of air saturation) is normally beneficial for keeping a high cell viability and thus for extending the production phase, it may become limiting for cellular metabolism, leading to conditions of hypoxia. Under these conditions, certain cellular functions such as mitochondrial activity can be impaired. The mitochondrial activity has been reported to strongly depend on the oxygen level in the range of 2–10% of air saturation for different cell lines. The declined activity of mitochondria can reduce the specific secretion rate of proteins and also the volumetric accumulation of product in most cases. Thus, although hypoxia is considered beneficial in delaying cell death and minimizing chromosomal damage, these effects may not be as critical as ensuring an appreciate delivery of oxygen to cells in a bioreactor to avoid starvation or low product yield. Animal cells used for protein production should therefore be cultivated within an optimal range of DO. The optimum oxygen concentration for cell growth varies with cell type and has been reported to be in the range of 10–50% of air saturation. For the formation of antibodies by hybridomas a DO range between 20 and 70% of air saturation has been reported to be optimal for the volumetric productivity.

Under optimal oxygen supply the specific oxygen consumption rate of most cultured animal cells ranges from 0.15 to 0.8×10^9 mmol O^2/cell/hr. The volumetric oxygen demand of 0.1–1.0 mmol O^2/L/hr at a typical cell concentration of 2×10^6 cells/mL is much lower than the oxygen demand (13–199 mmol O^2/L/hr) observed for mold and yeast cultures. However, unlike microorganisms, animal cells do not have cell wall and cannot be stirred vigorously in culture. Although hybridomas have been successfully grown in airlift reactors, it has been shown that mammalian cells can be damaged by gas bubbles used to increase the gas–liquid interfacial area in sparged reactors. An increase in oxygen demand will result from the use of perfusion cultures or fed-batch cultures that can increase the cell concentration by 10–50 times. With the development of high cell density culture aeration strategies that can meet the high oxygen demand while efficiently removing carbon dioxide and minimizing bubble damage to the cells have become an important issue.

The rate of oxygen uptake can be estimated from the consumption rates of nutrients. Stoichiometrially, the specific oxygen consumption rate (qO_2) of mammalian cells can be written as :

$$qO_2 = \alpha\left(q_{\mathrm{Glc}} - \frac{1}{2}q_{\mathrm{Lac}}\right) + \beta q_{\mathrm{TAA}} \qquad \text{...(29)}$$

where q_{TAA} is the specific consumption rate of total amino acids. α and β are stoichiometric coefficients representing the consumption of oxygen due to oxidation of glucose and amino acid in the energy metabolism. Originally, α is assumed to be 6 mol/mol based on the assumption that glucose that is not converted into lactate is completely oxidized in the tricarboxylic acid cycle. An examination of experimental data from three hybridoma cell cultures reveals that this assumption leads to a unrealistic negative value for β. Obviously, glucose that is not converted to lactate is not completely oxidized. α should be therefore estimated from experimental data. α appears to be not cell line specific for hybridomas culture, whereas the value of β should be individually estimated for each culture. Equation (29) described the experimental data satisfactorily. It is interesting to note that α is only about half of the theoretical value assumed for complete oxidation of glucose. The two cultures, which were grown on media with serum and supplements rich in amino acids, respectively, have also almost the same β value. The culture that was grown on a serum-free medium has a much lower β value, suggesting that a greater portion of the amino acids is used for biosynthesis in this case. Under conditions of glutamine excess, Eq. (29) can be reduced to:

$$\begin{aligned} qO_2 &= \alpha\left(q_{\mathrm{Glc}} - \frac{1}{2}q_{\mathrm{Lac}}\right) + \beta\frac{q_{\mathrm{TAA}}}{q_{\mathrm{Gln}}}q_{\mathrm{Gln}} \\ &= \alpha\left(q_{\mathrm{Glc}} - \frac{1}{2}q_{\mathrm{Lac}}\right) + \beta^{*}q_{\mathrm{Gln}} \qquad \text{...(30)} \end{aligned}$$

where β* is a constant that is about 1.6 times as high as β.

A useful parameter for online control of animal cell culture is the ratio of consumption rates of oxygen (OUR) to glucose (Q_{Glc}). The ratio OUR/Glc was used to control the feeding rate of glucose and other nutrient components so as to keep the glucose concentration at a relatively low level. However, the latter changes during the cultivation.

This variation of the stoichiometric ratio can be described by a modified form of Eq. (30):

$$\frac{\text{OUR}}{Q_{\text{Glc}}}=\frac{qO_2}{q_{\text{Glc}}}=\alpha\left(1-\frac{1}{2}\frac{q_{\text{Lac}}}{q_{\text{Glc}}}\right)+\beta\frac{q_{\text{TAA}}}{q_{\text{Gln}}}\frac{q_{\text{Gln}}}{q_{\text{Glc}}} \quad ...(31)$$

The rate equations presented above for the product formation and consumption of nutrients may be directly used to predict the ratio of oxygen and glucose consumptions as functions of residual concentrations of glucose and glutamine and the cell growth rate. The ratio of consumption of total amino acids to glutamine can be estimated as:

$$\frac{q_{\text{TAA}}}{q_{\text{Gln}}}=\text{TAA}_{\min}e^{A/C_{\text{Gln}}} \quad ...(32)$$

where $\text{TAA}_{\min}$ is the minimum ratio of consumption of total amino acids (including glutamine) to glutamine and typically has a value about 1.6 mol/mol for most cell cultures. *A* is a constant and has a typical value of 0.01–0.02. Similarly, ratio equations for $q_{\text{Lac}}/q_{\text{Glc}}$ and $q_{\text{Gln}}/q_{\text{Glc}}$ have been derived that can be used to convert Eq. (31) into the following simplified functional form:

$$\frac{\text{OUR}}{Q_{\text{Glc}}}=\frac{qO_2}{q_{\text{Glc}}}=\alpha\left(1-\frac{\text{LG}_{\max}}{2}\cdot\frac{C_{\text{Glc}}}{C_{\text{Glc}}+K_{\text{LG}}}\right)$$
$$+\beta\cdot\text{TAA}_{\min}\cdot e^{A/C_{\text{Gln}}}\cdot\text{GG}_{\min}\cdot e^{B/C_{\text{Glc}}} \quad ...(33)$$

In Eq. (33), $\text{LG}_{\max}$ is the maximum lactate yield from glucose, having a value of about 1.5–1.7 mol/mol for most cell lines; K_{LG} is a saturation constant reflecting the sensitivity of lactate formation to the residual concentration of glucose, and hasvalues between 0.1 and 0.3 mmol/L for most cell lines; $\text{GG}_{\min}$ is the minimum ratio of consumptions of glutamine to glucose. It typically has values between 0.2 and 0.4 mol/mol. *B* is a constant that depends on the cell lines. For hybridoma cells *B* has a typical value of 0.03–0.05 mM. Equations (31) and (33) can be used to predict the upper and low limits for OUR/Glc and the influence of nutrient availability.

Xiu et al. presented a method to estimate the rates of oxygen uptake and carbon dioxide evolution of animal cells based on material and energy balances. For this purpose, lumped compositions, molecular weight, and reductance degree of cellular proteins, MAb, biomass, and amino acid consumption (excluding glutamine and alanine) are used, which were found to be relatively constant for different hybridoma cell lines and can be used as regularities. The calculated rates of oxygen uptake and carbon dioxide evolution agreed well with experimental values of several different cultures reported in the literature. The method of Xiu et al. gives comparable results as calculated with Eq. (36) or on the basis of a detailed metabolic reaction network.

Dissolved Carbon Dioxide

Carbon dioxide is produced via catabolic reactions and is required for the synthesis of pyrimidines, purines, and fatty acids in animal cells. CO_2 produced by cells enters the culture broth in a dissolved form. In aqueous solution, CO_2 forms carbonic acid, which dissociates to bicarbonate and carbonate ions. The reactions involved can be written as:

$$CO_2\,(gas) \Leftrightarrow CO_2(aq.) + H_2O \Leftrightarrow H_2CO_3 \Leftrightarrow H^+ + HCO_3^- \Leftrightarrow 2H^+ + CO_3^{2-} \quad \text{...(34)}$$

The dissociation constants of H_2CO_3 (K_1) and HCO_3^- (K_2) can be written as:

$$K_1 = \frac{C_{HCO_3^-} C_{H^+}}{C_{CO_2} + C_{H_2CO_3}} \quad \text{...(35)}$$

$$K_2 = \frac{C_{CO_3^{2-}} C_{H^+}}{C_{HCO_3^-}} \quad \text{...(36)}$$

At 37°C, $pK_1 = 6.02$ and $pK_2 = 10.3$ according to Arrua et al.

If the concentration of CO_2 dissolved is in equilibrium with the gas phase, the average concentration of CO_2 ($[C_{CO_2}]^*$) throughout the liquid can be written as:

$$[CO_2]^* = H_{CO_2} P_{\chi_{CO_2}} \quad \text{...(37)}$$

where H_{CO_2} is Henry's constant for CO_2 [2.99×10^{-2} mM/mmHg at 37°C], P the total pressure of gas phase, and χ_{CO_2} the mol fraction of CO_2 in exit gas. $[C_{CO_2}]^*$ in Eq. (37) can be taken as the sum of C_{CO_2} (aq.) and $C_{H_2CO_3}$. Thus, from Eqs. (35) and (36) and according to the definition of pH,

$$[HCO_3^-] = \frac{[C_{CO_2}]^* K_1 10^{pH}}{[H^+]^{St}} \quad \text{...(38)}$$

$$[CO_3^{2-}] = \frac{[C_{CO_2}]^* K_1 K_2 10^{2pH}}{([H^+]^{St})^2} \quad \text{...(39)}$$

$[H^+]^{St}$ is the standard concentration of H^+, having the value of 1 mol/L. The concentration of CO_2 dissolved in the culture broths is in general higher than that calculated by assuming CO_2 in the broth to be in equilibrium with CO_2 in the gas phase, since a pressure gradient is required for CO_2 transfer from the liquid phase to the gas phase. The extent of this difference is dependent on the evolution rate of CO_2 and the mass transfer coefficient of the cultivation systems and can be calculated according to Zeng.

At pH values (7–7.2) normally used for animal cell cultures the decomposition of HCO_3^- to CO_2^{2-} can be ignored. By further assuming an effective equilibrium between CO_2 (aq.) and HCO_3^-, Eq. (34) can be simplified as:

$$CO_2(gas) \leftarrow\rightarrow CO_2(aq.) + H_2O \leftarrow\rightarrow H^+ + HCO_3^- \quad ...(40)$$

The equilibrium bicarbonate concentration ($[HCO_3^-]$; mM) at 37°C can be related to the partial pressure of CO_2 [$P\chi_{CO_2}$ in Eq. (37), expressed here as pCO_2 in mmHg] in the gas phase and the medium pH via the following equation:

$$\log[HCO_3] = pH + \log[pCO_2] - 7{:}543 \quad ...(41)$$

Thus, $[HCO_3^-]$ will increase with increasing pCO_2 and/or pH. It should be emphasized again that real concentration of $[HCO_3^-]$ in a culture system is higher than that calculated by the above equation and depends on the mass transfer of the system.

The increase in $[HCO_3^-]$ under elevated pCO_2, as well as the increase in cation concentration due the addition of a base such as NaOH to control pH, results in a concomitant increase in medium osmolality. High osmolality by itself may have detrimental effects on cell growth and metabolism ("Effect of pH"). It is therefore important to dissect the effects of dissolved CO_2 and osmolality. Many results reported in literature concerning the effects of dissolved CO_2 are in fact combined effects of CO_2 and osmolality. Caution should also be taken in comparing the effects of dissolved CO_2 in different culture systems due to the fact that the dissolved concentration of CO_2 and the corresponding concentration of $[HCO_3^-]$ can change significantly depending on the mass transfer characteristics of the culture system and the physiochemical properties of the culture medium. For example, CO_2 accumulation is high in cultures oxygenated with a low flow rate and small bubbles of pure O_2 where pCO_2 was predicted to reach 150–200 mmHg. pCO_2 values in this range have been reported for a 200-L culture sparged with small amounts of O_2, 1800–2500-L production bioreactors, and a high cell density perfusion bioreactor. The physiological pCO_2 range for proper growth of animal cells is 31–54 mmHg.

Carbon dioxide partial pressure in the range of 120–200 mmHg has been shown to inhibit growth and recombinant protein production and protein glycosylation. For example, Drapeau et al. showed that the growth rate and the cell-specific M-CSF production of CHO cells were both 40% lower at a pCO_2 of 165 mmHg than at 53 mmHg. With another CHO cell culture, Gray et al. found dose-dependent decreases in the cell density, viability, and specific production rate—such that the total productivity of a recombinant viral antigen in a 10-L perfusion bioreactor decreased by 69% at pCO_2 = 148 mmHg. Decreases in specific productivity of BHK-21 cells have been observed beginning at 50–80 mmHg. It has also been reported that recombinant protein production by infected Sf-9 insect cells is markedly delayed under 115mmHg pCO_2. Early cell death has been observed in NS/0 myeloma cell cultures with a final pCO_2 of 120 mmHg.

Due to the formation of bicarbonate, an increase in pCO_2 at a constant pH will result in a proportional increase in osmolality. deZengotital et al. demonstrated that the growth rate of a hybridoma AB2-143.2 cell line in well-plate culture decreased with increasing pCO_2, with a 45% decrease at 195 mmHg pCO_2 under a partial osmolality compensation (to 361 mOsm/kg). Inhibition was more extensive without osmolality compensation, with a 63% decrease in growth rate at 195 mmHg pCO_2 and 415 mOsm/kg. Also, the death rate of the hybridoma cells increased with increasing pCO_2, with 31- and 64-fold increases at 250 mmHg pCO_2 for osmolality at 401 and 469 mOsm/kg, respectively. The specific glucose consumption and lactate production rates were 40–50% lower at 140 mmHg pCO_2. However, there was little further inhibition of glycolysis at higher pCO_2. But the specific antibody production rate was not significantly affected by pCO_2 or osmolality within the range tested. Interestingly, quite different results were obtained with the same hybridoma cell line in continuous culture. For example, the death rate kd decreased only slightly at 140 mmHg in continuous culture, while in well-plate and batch cultures kd was more than twice as great as that for the control at 140mmHg. The authors explained this difference by the high residual nutrients and low byproduct levels in the continuous culture. It is also possible that the different concentrations of dissolved CO_2 and thus HCO_3^- in the different culture systems also contributed to the different observations.

Osmolality and Salt

Osmolality is one of the most important physical factors in mammalian cell cultures. Most cell culture media are designed to have an osmolality in the range of 270–330 mOsm/kg, which is known to be acceptable for most cells. Osmolality has been shown to affect both cell growth and protein production. The impact of osmolality on cell growth is cell-type specific, with CHO cells exhibiting less growth inhibition than hybridoma cells exposed to similar medium osmolality. Osmolality compensation by decreasing the concentration of NaCl in the basal medium has been found to partially mitigate the inhibitory effect of elevated pCO_2 on cell growth and protein production and on neural cell adhesion molecule polysialylation by CHO cells. Additionally, HCO_3^- free perfusion medium has been used to greatly decrease the build-up of pCO_2 in perfusion culture.

As mentioned earlier, dissolved CO_2 often interferes or even masks the effects of osmolality. Recently, deZengotita et al. performed a thorough study to decouple the effects of pCO_2 and osmolality on the growth and metabolism of hybridoma cell by using low-salt basal media and corresponding compensation of osmolality. Under control conditions (40 mmHg; 320 mOsm/kg), cell growth and metabolism was similar in DMEM: F12 with 2% fetal bovine serum and serum-free HB GRO. In both media, pCO_2 and osmolality made dose-dependent contributions to the inhibition of hybridoma cell growth and synergized to more extensively inhibit growth when combined. Elevated osmolality was associated with increased apoptosis.

Specific antibody production also increased with osmolality although not with pCO_2. On the other hand, osmolality had little effect on glycolysis while elevated pCO_2 (with or without osmolality compensation) inhibited glycolysis in a dosedependent fashion in both media. Mammalian cell lines derived from different organs have very distinct functions and therefore very different enzymatic and metabolic patterns. They may also have different responses to osmotic stress. For example, AB2-143.2 cells increase the specific formation rate of MAb (q_{MAb}) in response to gradual osmotic stress in continuous culture. In contrast, IND1 cells decrease q_{MAb} during gradual osmotic stress. Although the two cell lines differ in antibody production, they have very similar intracellular antibody content profiles. Both cell lines show: (1) a constant antibody content during the exponential phase in control culture with a decrease as cells enter the stationary phase; (2) maintenance of exponential-phase antibody content into the stationary phase after batch osmotic shock; and (3) no change in antibody content in response to gradual osmotic stress.

Cell culture longevity and thus product formation in fed-batch culture of hybridomas is often limited by elevated medium osmolality caused by repeated nutrient feeding. The use of hypoosmolar medium can overcome some of the problems cause by osmolality. As shown by Ryu and Lee , the use of hypoosmolar medium (223 mOsm/kg) as an initial medium in fed-batch culture can delay the onset of severe cell death of hybridoma cells, resulting in improved cell longevity and a substantial increase in the final antibody concentration.

Rate Laws of Cell Growth and Death

In most unstructured models, the specific growth rate μ is often expressed as a function of the glucose and/or glutamine concentration. In some of the models inhibition terms for lactate and ammonia are also included. Similar models are also used for the death rate (k_d). Portner and Schafer presented a survey of the unstructured models for cell growth and death rates. They also made a quantitative comparison between selected unstructured growth models and experimental data. It was noticed that a very limited data set covering a relatively narrow range of experimental conditions has often been used for model set-up. This practice of model formulation can lead to incorrect conclusions and model parameters that are not realistic. For example, inhibition constants estimated for lactate and ammonia considerably differ from those experimentally separately determined; saturation constants estimated for glucose ranges from as low as 0.03 mM to as high as 1mM for similar hybridoma cell lines.

Zeng et al. analyzed experimental data from six hybridoma cell lines grown on serum-containing and serum-free media in both normal continuous and perfusion cultures with respect to the significance of nutrients and products in determining the growth and death rates of cells and with respect to their mathematical description. It was interesting to find that for many continuous cultures there seems to exist direct correlations between μ and certain culture variables. For example, if one plots μ versus the residual glutamine concentration of a continuous hybridoma culture studied by Hiller et al., an apparent strong dependency of μ on C_{Gln} is observed. The growth rate curve appears to follow a Monod-type saturation kinetic that would give an apparent saturation constant as low as 0.01–0.05 mM for glutamine. Hiller et al. studied the same cell line in perfusion culture with glucose and glutamine concentrations in medium different from those used in the normal continuous culture. Interestingly, the growth rate of the perfusion culture apparently shows similar correlation with C_{Gln}, following again a kind of saturation kinetic with a saturation constant as high as 1.5–2 mM and a minimum C_{Gln} of about 1.2 mM for growth. Similar apparent dependencies can also be demonstrated for more cultures in the literature. It is worth mentioning that the continuous culture of Frame and Hu was clearly shown to be glucose-limited for some of the steady states. The possibility of a glutamine limitation was experimentally excluded. An apparent K_m value of about 1 mM would be estimated from these data if the culture were assumed to be glutamine-limited. The medium used for the latter culture had an intermediate concentration of glutamine (3.61 mM) compared to 2.5 mM for the chemostat culture of Hiller et al. and 6 mM for the perfusion culture. It is thus expected that the apparent saturation curve for the cell line of Hiller et al. would move toward the left side if lower initial glutamine concentrations were used in the medium for perfusion. By carrying out a series of continuous cultures at varied initial glutamine concentration any apparent K_m values for glutamine between 0.01 and 2.5 mM would be obtained for one and the same cell line under similar environmental conditions. Obviously, an assumption of glutamine limitation for these cultures would be very implausible. It should be emphasized, however, that such a controversial conjecture is only obvious if one compares experimental data obtained by independent changes of culture conditions such as the medium composition and the ratio of D_B to D. In conclusion, the apparent strong dependency of μ on C_{Gln} may not necessarily mean a glutamine limitation. It can be just coincidental. In analogy, the apparent correlation of m with glucose for some of the cultures studied in literature may not necessarily mean a glucose limitation as well.

Ammonium and lactic acid are often considered to be the dominating inhibitors limiting growth as they are the main byproducts of the substrate metabolism and have been shown to be toxic to cell growth above certain concentrations. Plotting μ versus ammonium and lactate concentrations for the cultures of Hiller et al. As in the case of glutamine a certain degree of correlation is observed between m and the ammonium concentration. Again, the correlation depends strongly on the operation mode and obviously on the initial concentration of glutamine in the medium. There is no clear correlation between μ and the lactate concentration. It can be stated that for most cell cultures lactate and ammonium cannot be the dominant factors determining growth rate. The possibility of growth inhibition by the product MAb was also examined. In a similar way as argued for ammonium, MAb produced appears not to be a dominant inhibitor in the lines examined.

Three conclusions can be drawn regarding the significance of nutrient limitation and product inhibition and the modeling of cell growth rate. First, in all the cultures examined neither the macro-nutrients such as glucose, glutamine, and other amino acids nor the three products lactate, ammonium, and MAb are clearly dominant factors affecting the growth rate. Second, the apparent correlation of μ with some of these variables can be simply coincidental due to stoichiometric and/or kinetic interdependencies of the variables if they are not independently varied. Finally, none of the unstructured growth models existing in the literature can be generally applied, especially if they are intended to cover a relatively wide range of experimental conditions. Similar analysis and conclusions can be made with respect to the significance of nutrients and products in affecting the cell death rate. For seven continuous cultures it was then found that μ almost linearly correlates with the ratio of the viable cell concentration (N_v) to the dilution (or perfusion) rate (D) irrespective of the operation mode. Similarly, the specific death rate (k_d) of all the cultures is a function of the ratio of the total cell concentration (N_t) to the dilution (perfusion) rate. For most of the cultures k_d also linearly correlates with μ if the effect of N_t/D is taken into account.

Based on these observations and other experimental evidence that suggests the formation of a not yet identified critical factor or autoinhibitor by the cells the following simple rate equation was derived for the specific growth rate of hybridoma cells:

$$\mu = \mu_{max}\left(1 - \frac{C_I}{C_I^*}\right)\frac{C_{Glc}}{C_{Glc} + K_{Glc}}\frac{C_{Gln}}{C_{Gln} + K_{Gln}}\frac{K_{Lac}}{C_{Lac} + K_{Lac}} \times \frac{K_{NH_3}}{C_{NH_3} + K_{NH_3}} \quad ...(42a)$$

where C_I is the concentration of the autoinhibitor, C_I^* is the maximum concentration of the autoinhibitor above which cells cease to grow. Assuming a specific formation rate q_I for this inhibitor by viable cells, Eq. 42a(a) can be transformed to:

$$\mu = \mu_{max}\left(1 - \alpha\frac{N_v}{D}\right)\frac{C_{Glc}}{G_{Glc} + K_{Glc}}\frac{C_{Gln}}{C_{Gln} + K_{Gln}}\frac{K_{Lac}}{C_{Lac} + K_{Lac}}$$

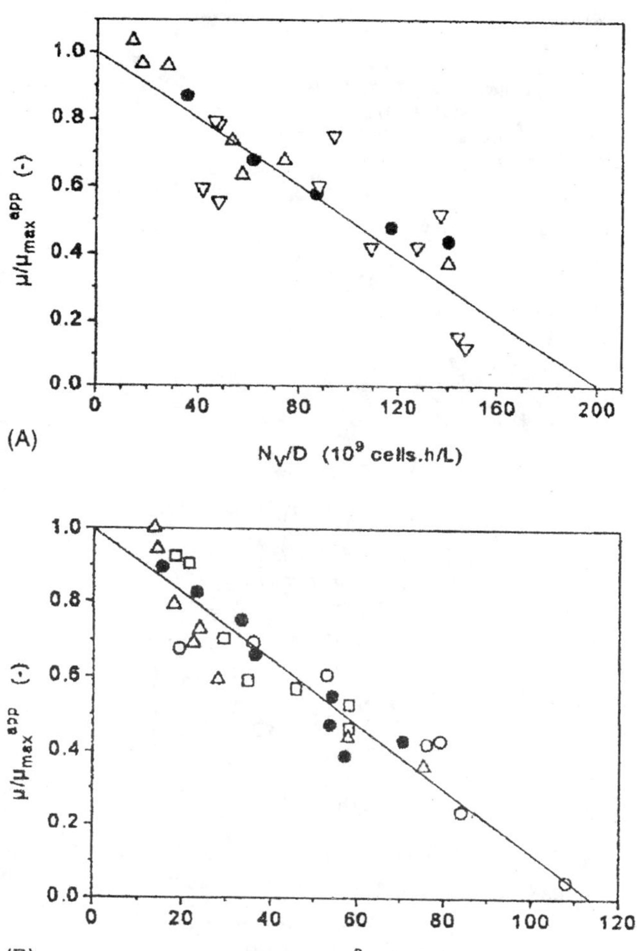

Fig. 4.3. Relationship between μ/μ^{app} and the reciprocal of the specific dilution rate N_v/D for cells grown on (A) serum-containing media and (B) serum-free media.

$$\times \frac{K_{NH_3}}{C_{NH_3} + K_{NH_3}} \qquad \ldots(42b)$$

for continuous culture, with the constant

$$\alpha = \frac{q_I}{C_I^*} \qquad \ldots(43)$$

Similarly, the following rate equation has been proposed for the specific death rate in continuous culture:

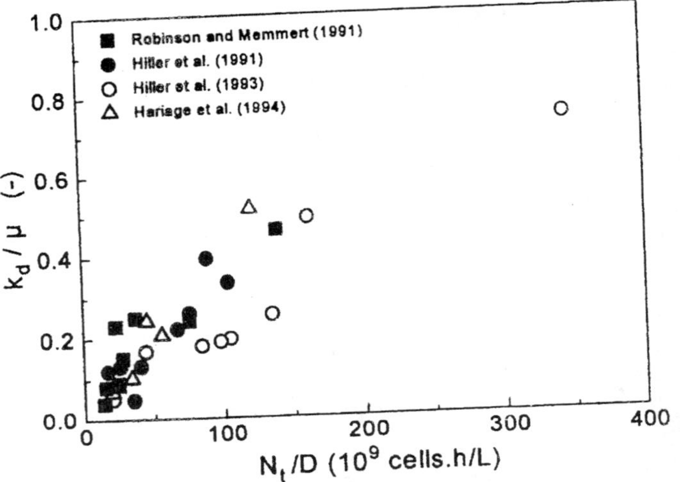

Fig. 4.4. Correlations between k_d/μ and N_t/D for hybridoma cells grown in different continuous cultures with serum-free media.

$$k_{\rm d} = (\beta_0 + \beta\mu)\frac{N_{\rm t}}{D} \qquad \ldots(44)$$

where β_0 and β are constants.

In most cases, Eq. (42) can be simplified as:

$$\mu = \mu_{\rm max}^{\rm app}\left(1 - \alpha\frac{N_{\rm v}}{D}\right) \qquad \ldots(45)$$

where $\mu_{\rm max}^{\rm app} = f$ (nutrients, lactate, ammonia, etc.) $\approx$ constant is an apparent maximum specific growth rate that includes the influences of other inhibitors such as ammonium and lactate. The physiological meaning of α can be easily understood by setting μ=0 in Eq. (45). Under these conditions,

$$\alpha = \left(\frac{D}{N_{\rm v}}\right)_{\rm min} \qquad \ldots(46)$$

According to Eq. (46), α has the unit "liter per unit biomass per hour" and can be defined as the minimum specific dilution (or perfusion rate (based on viable cells) of a continuous culture to keep a non negative cell growth rate. For perfusion culture the perfusion rate should replace the dilution rate D in Eqs. (42)–(45).

Equations (44) and (45) describe the experimental data of seve continuous cultures reasonably well. To calculate the growth rate of batch culture the original form of Eq. (42a) should be used. Equatio

(44) should be modified for batch culture. It is worth mentioning that irrespective of the cell lines, cells grown on serum-containing media have almost the same α value that is distinctively different from that of cells grown on serum-free media. This indicates that under the current cultivation conditions the formation rate of the autoinhibitor(s) or the sensitivity of cell growth and death to the autoinhibitor(s) is mainly affected by the medium composition; α can be used as a quantitative parameter for comparison of medium performance.

The positive relationship between μ and k_d as envisaged by Eq. (44) is in strict contrast with the prevailing view in the literature that k_d is inversely related to μ. The apparent negative correlation between k_d and μ found for some normal continuous cultures appears to be coincidental due to the stoichiometric and kinetic interdependencies of culture variables as discussed earlier for the correlations of μ with concentrations of nutrients and products. In fact, the inverse relationship between k_d and μ is often not valid for perfusion cultures. The apparent inverse relationship between k_d and μ in a normal continuous culture can be understood with the help of Eq. (44). At low dilution rates (and thus low values of μ), the ratio $\mu(N_t/D)$ is normally higher than that at high dilution rates. Hence, k_d is higher at lower growth rates. But this is by no means an intrinsic relationship. This is also supported by the experimental results of Vomastek and Franek. They studied the kinetics of the development of apoptosis in mouse B-cell hybridoma cultures grown on both protein-free medium and serum-containing medium. The B-cell hybridoma apoptosis was found to be associated with cell proliferation and metabolic activity.

Product Formation: Monoclonal Antibody

Biologically, the rate of synthesis and secretion of MAbs depends on the rate of peptide chain synthesis, chain assembly, interorganelle transport, intravesicular degradation, and release from the cell membrane. The genetic make-up of a particular cell line may dictate these events intracellularly. However, the rate of these individual steps and the overall rate of MAb production can also be greatly affected by cell growth and culture environmental conditions. These include growth rate, cell division cycle and viability, components of culture medium or broth, cultivation conditions, reactor type and operation mode, and hydrodynamic stress. It is not the purpose of this chapter to assess all of these factors. Instead, only those that are adjustable for a given cultivation system (i.e., for given reactor type, operation mode, and hydrodynamic parameters) and relevant for kinetic modeling are mentioned. The effects of cell cycle, growth rate, and viability have received the most attention in literature. Both synthesis and secretion are shown to be maximum in the G1/early S phases. It seems to be generally accepted that MAb is mainly produced by secretion of viable cells, though some experimental results showed that high death rate of cells can apparently cause increase in the specific MAb production rate (q_{MAb}). In many cell lines q_{MAb} was found to be higher at lower growth rate. These results are consistent with the finding that a prolongation of the G1-phase and an increase in death rate increased q_{MAb} of different cell lines. In addition to the effect of cell cycle it is understood that an elevated growth rate may make a significant portion of metabolic energy unavailable for the synthesis of antibody, leading thus to a decreased MAb synthesis rate.

In most studies the G1-phase arrest of cell cycle and the reduction of growth rate were achieved by starvation of cells for an essential nutrient or energy source or by the addition of DNA synthesis inhibitors such as thymidine and hydroxyurea, or genotoxic agents such as adriamycin. These approaches are not adequate to achieve a prolonged production phase with high productivity since they interfere with cell viability and/or disturb metabolic processes necessary for protein synthesis. Regulation of cell growth rate and cell cycle through the use of fed-batch culture with careful nutrient feeding was shown to be feasible with a murine myeloma cell line. Recently, the use of conditional overexpression of tumor suppressor genes such as p21, p27, or p53 (*cytostatic*) to regulate cell growth rate has received increasing interest. The MAb productivity of myeloma and CHO cell lines could be improved by about 3–4-folds through cytostatic gene overexpression and cell growth arrest at the G1-phase of the cell cycle. Preliminary study into the cytostatic function and its relationship to productivity demonstrated that the arrest of CHO cells by p21 overexpression correlates with the cellular response to accumulation of mitochondrial mass, ribosomal protein S6, and intracellular protein uncoupling cell growth.

By combining the cell cycle theory with the estimated number of MAb-coded mRNA molecules per cell and by considering the relative time length of the individual phases, Suzuki and Ollis developed a structured model that predicted an enhanced MAb production at low growth rate. In contrast to this theoretical analysis and most of the experimental reports, Robinson and Memmert showed with a transfactant of a myeloma (SP2/O) cell line grown in a serum-free medium that q_{MAb} increased linearly with m. Borth et al. also showed with a human–mouse heterohybridoma (3D6-LC4) that q_{MAb} increased with μ. It was also reported that MAb production is not growth associated in some cell lines. Ray et al. reported on an optimal μ for q_{MAb}. It is not clear if these differences are due to the inherent characteristics of different cell lines or to some extent due to the very different experimental conditions that may significantly affect MAb formation as well.

A major concern about the use of continuous culture for MAb production is the possible loss of culture productivity. It has been reported for several cell lines that hybridomas may dramatically lose their MAb productivity soon after revival from liquid nitrogen storage (thawing). This loss of productivity is not reversible. It may last for a few weeks or months for cells to reach a stable MAb productivity in continuous culture depending on the cell line and/or culture conditions. The mechanisms for the loss of MAb productivity in hybridomas are not well understood. In a culture of mouse–mouse hybridoma cells the loss of MAb was shown to be mainly due to the occurrence of a nonproducing subpopulation of cells. Other mechanisms may involve the loss of heavy chain and/or light chain production and the loss of the chromosomes containing the gene loci for the antibody chains similar to those for myelomas. Whatever the mechanisms are the available literature data suggest that it is important to consider culture stability when studying the effects of cell growth and other environmental conditions on MAb production, particularly when comparing kinetic data of different cell lines or experimental data obtained in batch culture.

In a comprehensive study of hybridoma growth and product formation using a statistic experimental method, Gaetner and Dhurjati showed that the concentrations of base medium, glucose, glutamine, serum, lactate, and ammonium all have significant effects. q_{MAb} increased with increasing concentrations of base medium and serum but decreases with lactate and ammonium. The negative effects of lactate and ammonium at high concentrations have also been confirmed for other cell lines. Concerning the effect of glucose concentration it was shown to inversely affect the formation of MAb. This negative glucose effect seems to be consistent with the finding that the immunoglobulin heavy chain binding protein, which is involved in antibody assembly, is identical to the 78 kD glucose-regulated protein. The latter is induced by glucose starvation. In contrast, high glutamine concentration in the culture was found to stimulate MAb formation. Other factors affecting MAb production include temperature, pH, dissolved oxygen, vitamin, amino acids, osmolarity, and probably cell density. Whereas some of these factors (such as pH and pO_2) are constant or do not change very much during a cultivation and therefore may not be important for kinetic modeling, changes of other factors or the depletion of an essential nutrient may strongly alter the product formation pattern. It is thus important to identify possible stoichiometric and/or kinetic limitations when modeling the product formation.

Several structured models have been proposed to describe the intracellular synthesis and transport of MAbs. In contrast, few works dealt with the quantitative description of the influences of extracellular culture conditions. A review of these models was given by Tziampazis and Sambanis. In most of the models, merely variables associated with cell growth such as the specific growth rate, the death rate, and cell cycle, are considered. Some authors considered the effects of other factors such as serum, glutamine, lactate, and ammonium on q_{MAb} in mathematical terms. These expressions were, however, all derived for very specific experimental conditions and normally consider only one factor. Zeng made an attempt to mathematically describe the effects of culture conditions on MAb production by hybridoma cells, with particular emphasis on cultures under unsteady-state conditions. The following general rate equation that takes account of productivity loss during long-term cultivation, cell proliferation, and the effects of nutrients and toxic products was proposed:

$$q_{\mathrm{MAb}} = (\alpha + \beta k_{\mathrm{d}}) \prod_i \frac{C_i}{C_i + K_i} \prod_j \frac{K_j}{C_j - C_j^* + K_i} (B + e^{-A\Delta t}) \quad \ldots(47)$$

In Eq. (47), k_{d} is the specific death rate of cells. C_i and C_j are the concentration (or intensity) of a stimulating factor i and the concentration (or intensity) of an inhibiting factor j, respectively. α, β, K_i, and K_j are constants. The term $(B + e^{-A\Delta t})$ describes the loss of MAb productivity of cells during long-term cultivation.

For a reliable assessment of effects of different factors and for a comparison of kinetic data on MAb production it is important to consider possible loss of antibody productivity, the time dependence of which can be modeled by an exponential function plus a constant term $(B + e^{-A\Delta t})$. The latter gives an indication of stability of MAb productivity of a given cell line under the experimental conditions. Among the parameters related to cell proliferation the cell death rate appeared to be useful for modeling MAb production under both steady-state and unsteady-state conditions. Model analysis of the so-called cell density effect suggested that it can be attributed to the varying availability of nutrients. Among others, the relative concentration of glucose and glutamine may be decisive for the production of antibody under certain conditions. Furthermore, the specific formation rate of antibody in perfusion culture was found to be strongly affected by the perfusion rate, indicating that antibody production is limited by component(s) of the medium not yet identified.

Cell Density Effect

As mentioned earlier, cell density has been reported to be an important parameter for the kinetics of batch culture. For a continuous perfusion culture at relatively high density, Banik and Heath showed that the specific rates of glucose consumption, glutamine consumption, and lactate production decreased significantly with increasing cell density. Similar effects of cell density were also reported by other authors. In the work of Banik and Heath, the specific antibody production was also a strong function of cell density, increasing as cell density increased independent of growth rate. Similar results were reported for a mouse–human hybridoma grown in serum-containing medium. However, constant q_{MAb} values were obtained with the same cell line grown in serum-free medium over a density range of 3.5×10^5 to 7.6×10^6 mL^{-1} viable cells. Other investigators found decreased q_{MAb} values with increasing cell density.

Density-dependent apoptosis has been reported for several cell lines. Normally, this kind of apoptosis was observed under low cell density. It is thought to be due to the lack or reduction of survival signals provoked by direct cell-to-cell contact or soluble growth factors (autocrine factors) produced by the cells. High cell density can also induce apoptosis which may have an important physiological role in vivo such as for the prevention of tissue hyperoxia and maintenance of appropriate tissue volume.

The cell density effects on cell growth and death are mathematically described in the rate Equations (44) and (45) Little is known about the effect(s) of cell density on the metabolism and MAb production. In the literature mechanisms modulated by the so-called *humoral factors* and *autocrine growth factors* were proposed to describe cell density dependent growth. In most of the experimental studies, however, no factors other than cell density itself and/or physical environments could be identified as the cause(s) for the observed effects. In the work of

Zeng, the rate equations mentioned earlier (22), (23), (26), (27), and (47) were used to examine the significance of possible cell density effect(s) on two hybridoma cultures grown in perfusion bioreactors. By replacing the nutrient concentration C_s with a "specific concentration of nutrient c_s"

$$c_S = \frac{C_S}{N_v} \qquad ...(48)$$

Equations (22), (23), (26), (27), and (47) satisfactorily describe the consumption rates of glucose and glutamine and the formation rates of lactate, ammonium, and MAb over a viable cell density range of 10^6–10^7 mL^{-1}. A density effect can be generally assumed for these cultures. However, the effect is only signifi- cant for the rates of glucose consumption and lactate formation of one cell line under the experimental conditions. The observed variations of metabolism and antibody production rate at different cell density appear to be attributed to the varying availability of nutrients. Among others, the relative concentration of glucose and glutamine may be decisive for the uptake of glucose and glutamine and for the production of lactate, ammonium, and antibody under certain conditions.

Models for Simulation of Cell Culture

Experimental data from kinetic studies in batch and continuous cultures build the basis for deriving quantitative relationships (rate equations) that correlate the rates of cell growth, death and metabolism with culture and environmental conditions. This set of rate equations represents a kinetic model of the cellular process. Combined with balance or differential equations governing the system, these rate equations can be used to simulate the time courses of concentrations of cells, nutrients, and metabolites in the culture under different conditions. Kinetic models for biological processes can be generally classified into four categories. In nonsegregated and unstructured models the cell population is treated as "*one-component solute*" or as an "*average cell.*" Cell growth is considered to be "*balanced.*" In segregated models the heterogeneity of cell population is considered and in structured models the intracellular structure and processes of cells are taken into account.

Unstructured Models

The simplest and frequently used kinetic models for cell culture are nonsegregated and unstructured models. Although the formulation of this kind of models is of empiric and phenomenological nature, they are usually based on fundamental observations of biological processes and can be used to qualitatively and quantitatively describe many important features of cell culture such as the dependency of growth rate on a limiting nutrient or an inhibiting substance. Because of the simple form and limited model parameters unstructured models are particularly suitable for process control, optimization, and scale-up of industrial processes. A major drawback of unstructured models is that they cannot be used for prediction and are generally not suitable for dynamic simulation.

The rate equations presented earlier were used to simulate the behavior of perfusion cultures at high cell density. As an example, comparisons of model simulations and experimental results of a perfusion culture. Model simulations of steady-state behavior are in good agreement with experimental results under varied perfusion and cell bleed rates except for cultures with very low viability. The rate equations can be also used to simulate the time courses of batch and

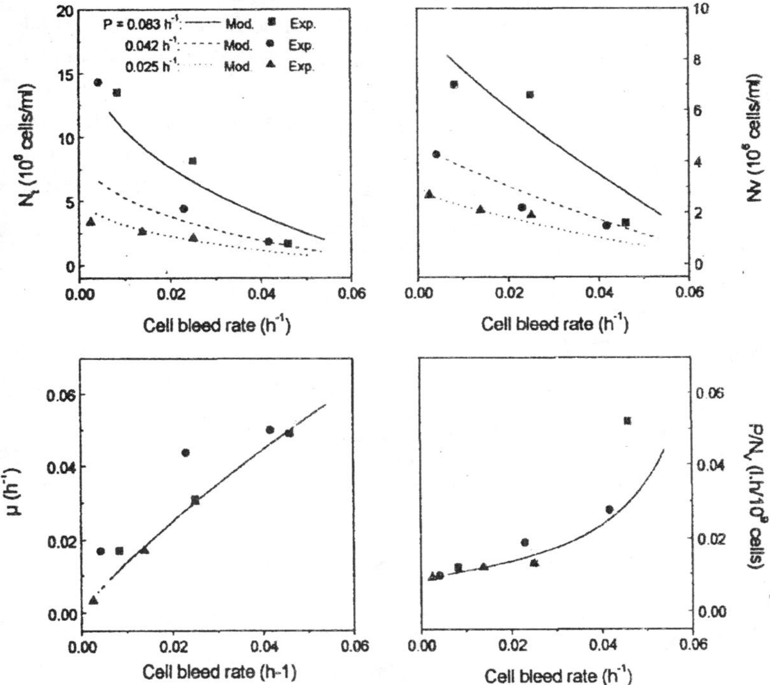

Fig. 4.5. Comparisons of model simulations and experimental results of total and viable cell concentrations, growth and specific perfusion rate of a perfusion culture with the cell line X-D at different perfusion and cell bleed rates.

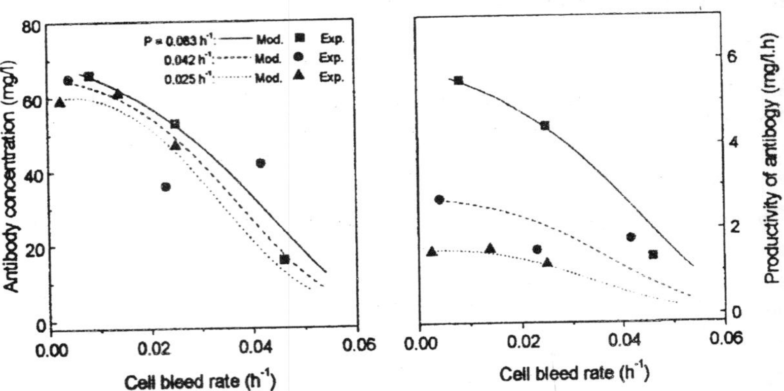

Fig. 4.6. Comparisons of model simulations and experimental results of concentration and productivity of monoclonal antibody of a perfusion culture with the cell line X-D at different perfusion and cell bleed rates.

continuous cultures under non-steady states. For example, the kinetic simulation of a culture that was first operated as a conventional continuous culture and then switched to perfusion. The effect of perfusion operation was described satisfactorily. It should be mentioned that under certain dynamic conditions such as sudden changes of nutrient supply the model may fail to describe the pitfall of cellular stress responses. The simulation capacities of the model appear to be mainly limited by the applicability of the cell growth and death rate expressions under these conditions. In general, unstructured kinetic models are severely

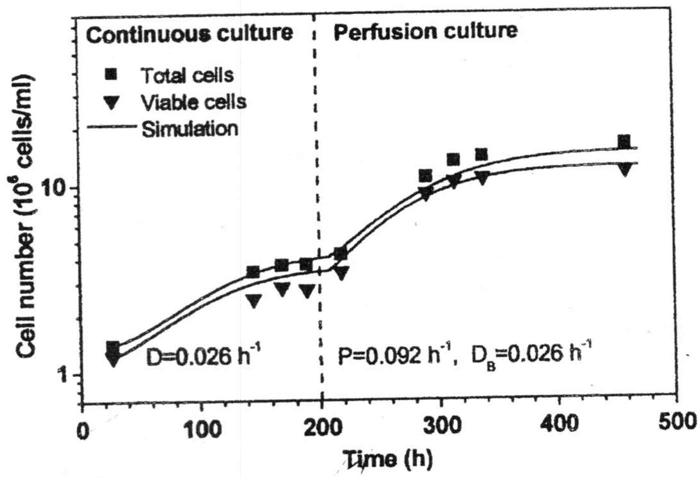

Fig. 4.7. Simulation of time course of growth of hybridoma cells in conventional continuous and perfusion cultures

limited for simulation of dynamic behavior of cell culture. To this end, structured models are preferred.

Structured Models

Structured or mechanistic models are based on knowledge of intracellular structure and bioreactions and their regulation mechanisms. One of the important advantages is their prediction ability. Once the model is experimentally verified, it can be in principle extrapolated for conditions outside the experimental range. It should be, however, mentioned that even the simplest living cell is so highly complicated system that any mathematical description is only an approximation of the real biological processes inside the cell. The underlying mechanisms used for model formulation are often not those that explain the biological processes on molecular basis but rather working hypothesis. Caution should thus be taken in interpreting the simulation results, especially for extrapolation. The experimental verification of a model does not necessarily guaranty the correctness of the hypothesis.

In the literature a number of structured or mechanistic models have been proposed for animal cell culture. Recently, cybernetic approach has also been used to model the kinetic behavior of cell culture, especially with respect to a possible occurrence of multiplicity in continuous culture. In the following, only two examples of these models are briefly introduced.

"Chemically or metabolically" structured models

In the chemically or metabolically structured models the biomass is divided into several chemical components (pools) such as cell membrane (lipids), protein, and nuclei acids. The stoichiometrical relationships (balance equations) and rate equations for these components build the basis for the modeling of the whole cell. Chemically or metabolically structured models can be generally formulated as follows:

$$\frac{dc_j}{dt} = \sum_{i}^{n} r_{ij} - \mu c_j \quad ...(49)$$

$$\mu = \hat{V} \sum_{i}^{n} \sum_{j}^{m} r_{ij} \quad ...(50)$$

where i stands for the n conversion reactions; j for the m components; c_j for the intracellular concentration of the component j; r_{ij} for the formation or consumption rate (based on intracellular volume) of the component j in the ith conversion processes including transport through

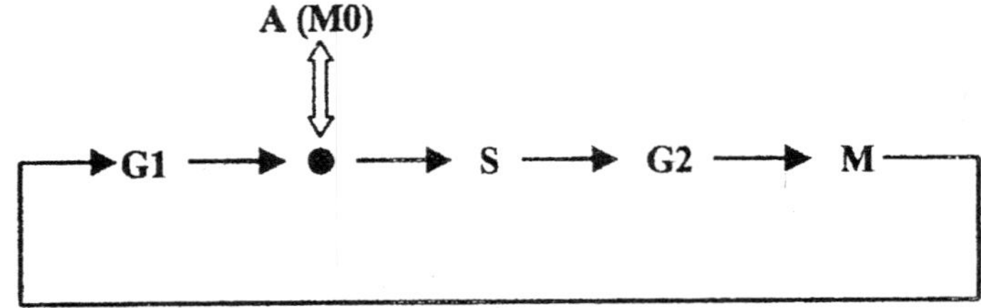

Fig. 4.8. The successive phases of cell cycle.

the cell membrane; $\hat{V}$ the intracellular volume per mass unit of biomass. The key and at the same time the difficulty in applying this modeling approach is the formulation of reliable kinetics rate expressions for r_{ij}.

The above model formulation was first practically used by Domach et al. and Domach and Shuler for the development of the so-called "*single-cell model*" for the growth of *Escherichia coli* on glucose. Batt and Kompala applied this modeling approach and developed a structure framework for simulating continuous cell culture. The cell is considered to be composed of four pools: amino acids, nucleotides, proteins, and lipids. The rate equations for these pools are mainly saturation type kinetic expressions. Interactions among the pools and the extracellular environment occur via nutrient consumption, metabolite formation, and transport of nutrients and metabolites.

A more promising structured model for cell culture is the extension of the above mentioned "*single-cell model*" for CHO cells by Wu et al. In the single-cell model for CHO, cells are considered to be composed of 18 component pools. The model can accommodate additional mechanistic details by further subdivision of any of the pools. Stoichiometry of pseudochemical reactions relating the mass or information flow of the pools is considered. The rate equations are also generally of saturation type with a feedback control term and have altogether more than 100 parameters. The effects of glucose, glutamine, other amino acids, and the cellular metabolic byproducts on cellular metabolism are considered in these rate equations. This model describes experimental data of CHO cells under both steady-state and transient conditions quite well. Wu et al. further extended this model to study the regulation of intracellular pH. It provides a direct mechanistic link between intracellular pH control and cellular metabolism.

Cell cycle model

In cell cycle models the biochemical features and kinetics of cells passing through the different phases of a cell cycle are considered.

The cell cycle is generally considered to consist of four essential phases: the gap 1 phase (G1-phase), the DNA synthesizing phase (S-phase), the gap 2 phase (G2-phase), and the mitotic phase (M-phase). After M-phase, which consists of nuclear division (mitosis) and cytoplasmic division (*cytokinesis*), the daughter cell enters a new cycle. A new cell cycle starts with the G1-phase, in which the cells, whose biosynthetic activities have been greatly slowed during mitosis, resume a high rate of biosynthesis. The S-phase begins when DNA synthesis starts and ends when DNA content of nucleus has doubled and chromosomes have replicated. The cell then enters the G2-phase, which ends when mitosis starts. Between the G1 and S phases a cell may stop dividing unless signaled to continue through the whole cycle. The cells stopping at late G1 is called an arrested cell and symbolized by "A". The arrested cells are also called to be in the G0 phase. Park and Ryu measured the population distribution of hybridoma cells during perfusion culture by flow-cytometric analysis. The results showed that, during the reduced growth period of perfusion culture, the fraction of cells in the S-phase decreased, and the fraction of cells in the G1/G0-phase increased with decreasing growth rate. The fraction of cells in the G2/M-phase was relatively constant during the whole period of perfusion culture. Ramirez and Mutharasan determined the cell cycle and mean cell volume distribution in batch culture by cytometric analysis. The mean cell volume closely correlated with the fraction of cells in the S-phase. On single cell basis, a cell generally increases in volume and mass from the beginning of the G1-phase through the S- and G2-phase until the end of the G2-phase.

Besides the morphological and biochemical differences cells at different cell cycle phases have remarkably different kinetic properties, particularly with respect to MAb productivity of hybridoma cells. Several investigations have shown that the maximum antibody synthesis rate occurs in the late G1-phase or early S-phase of hybridoma cells and appears to correlate with the faction of arrested cells. The morphological and biochemical differences of distinguished cell population during a cell cycle are the basis of cell cycle models. For example, the specific antibody production rate q_{MAb} can be represented as a summation of the fraction of cells in each of the five phases multiplied by the antibody production rate of a cell in that phase:

$$q_{\mathrm{Mab}} = k_{\mathrm{G1}} f_{\mathrm{G1}} + k_{\mathrm{S}} f_{\mathrm{S}} + k_{\mathrm{G2}} f_{\mathrm{G2}} + k_{\mathrm{M}} f_{\mathrm{M}} + k_{\mathrm{A}} f_{\mathrm{A}} \qquad \ldots(51)$$

where k_{G1}, k_{S}, k_{G2}, and k_{M} are the averaged antibody production rates in the G1-, S-, G2-, and M-phase, respectively; k_{A} is the antibody production rate of a cell arrested in the G1-phase; f_{G1}, f_S, f_{G2}, and f_M are the fractions of cells traveling the G1-, S-, G2-, and M-phase, respectively; and f_A is the fraction of cells arrested in the G1-phase.

The frequency of cells located at time t $(0 \le t \le t_C)$ measured from the beginning of the G1-phase is proportional to $2^{1-t/tc}$, where the total cell cycle time t_C is the sum of t_{G1}, t_S, t_{G2}, and t_M. t_{G1} includes the time of a cell arrested in the G1-phase and normally increases as the specific growth rate decreases, while t_S, t_{G2}, and t_M are approximately constant independent of the growth rate. Thus, t_C also increases as μ decreases.

The fractions of cells traversing the individual phases can be calculated as:

$$f_{G1} = (1 = f_A)\int_0^{x_{G1}} 2^{1-x}\frac{dx}{A} \qquad ...(52)$$

$$f_S = (1 = f_A)\int_{x_{G1}}^{x_S} 2^{1-x}\frac{dx}{A} \qquad ...(53)$$

$$f_{G2} = (1 = f_A)\int_{x_S}^{x_{G2}} 2^{1-x}\frac{dx}{A} \qquad ...(54)$$

$$f_M = (1 = f_A)\int_{x_{G2}}^{1} 2^{1-x}\frac{dx}{A} \qquad ...(55)$$

The fraction f_A is related to the specific growth rate μ and the cell cycle time t_C as follows:

$$f_A = 1 - \mu \cdot t_C / \ln 2 \qquad ...(56)$$

In a similar way, specific rates for growth and death, nutrient consumption, and formation of other metabolites may be related to the cell cycle. For example, in the work of Linardos et al. the death rate of the hybridoma is assumed proportional to the fraction of cells arrested in the G0-phase.

An underlying assumption in cell cycle models is that the rates of biosynthesis and degradation are constant in each cycle phase; the culture conditions affect only the viable cell number and the distribution of cells among the phases. This assumption may be only valid for certain cell systems under particular conditions.

Application of Models for Kinetic Analysis

As demonstrated earlier, mathematic models (rate equations) are useful for kinetic analysis of effects and interactions of different factors on cell growth and metabolism. They are particularly helpful for obtaining quantitative knowledge about the significance of these factors and their range of concentrations or values in which they are important. Models are also valuable for kinetic analysis of the overall performance and for the control of cell culture. This is briefly illustrated in the following for the kinetic analysis of perfusion culture. As evidenced, kinetic analysis of simulations and experimental results indicates that in perfusion cultures with a complete cell separation cell bleed rate is a key parameter that strongly affects all the process variables whereas the perfusion rate mainly affects the total and viable cell concentrations and the volumetric productivity of MAb. Growth rate, viability, and specific perfusion rate of cells are only a function of the cell bleed rate. This also applies to cultures with partial cell separation in the permeate if the effective cell bleed rate [Eq.(21)] is used. From the kinetic analysis it is clear that the (effective) cell bleed rate of a perfusion culture should be carefully chosen and controlled separately from the perfusion rate. In general, a low cell bleed rate that warrants reasonable cell viability is desirable for the production of antibodies. Furthermore, model simulations indicate the existence of an optimum initial glucose concentration in the feed. For several cell lines considered, the initial glucose concentration used in normal cell culture media is obviously too high. The initial glutamine concentration can also be reduced to a certain extent without significantly impairing the growth and antibody production but considerably reducing the ammonium concentration. The mathematical model can be used to predict these optimum conditions and may also be used for process design.

In this connection, it is worth mentioning that the specific perfusion rate based on viable cells (P/N_v) was proposed as a key control parameter for process scale-up and product consistency of perfusion cultures. The rationale behind this strategy can be understood in view of the linear relationship between μ and N_v/P [Eqs.(45)] under conditions of no obvious limitation of nutrients as discussed. The relationship between P/N_v and D_B is examined for varied perfusion rates. Surprisingly, P/N_v turned out to be almost only a function of D_B. The simple relationship between P/N_v and D_B underlines that P/N_v is a useful control parameter since it determines D_B which in turn flxes the cell viability, cell growth, and death rates. However, the so-called "*high-gain*" range for a P/N_v control algorithm corresponds to high cell bleed rates or high perfusion rate in systems with partial cell separation. From a practical point of view, the effective cell bleed rate seems to be a more direct and more favorable control parameter than P/N_v, at least for systems with a complete cell separation. In these systems, the desired cell bleed rate may be chosen based on considerations of cell concentration and physiology (e.g., viability, growth and death rates) and final product concentration. As mentioned earlier, a cell bleed rate as low as possible should be chosen for achieving high concentrations of cells and product. At such low values, D_B is quite sensitive to P/N_v. Thus, a small fluctuation in P/N_v may lead to relatively large changes of D_B and other process variables. This may be of disadvantage for process stability and consistency. Furthermore, an online measurement of viable cell concentration is not an easy task and is subjected to fluctuations. On the other hand, the control of D_B is simple and reliable. An online measurement of viable cell concentration is not needed. Once the D_B value is chosen the desired concentrations of cells and product can be achieved by varying the perfusion rate.

In general, simulations of cell culture, in particular detailed, structured, and dynamic simulations have a wide range of applications such as in planning and design of experiments, testing and development of theories and hypothesis concerning metabolism, data reconciliation, and inference of difficult to measure variables and advanced process optimization and control.

Concluding Remarks

Our knowledge about the kinetics of cell culture has been considerably augmented in recent years. Both unstructured and structured models are available to simulate the kinetic behavior of several cell lines in "*balanced*" batch growth or steady states of continuous cultures. These models are useful for quantitatively analyzing the influences of different culture and reactor operating conditions and for process control and optimization. However, difficulties are still encountered in simulating the dynamic behavior of cell culture under strongly disturbed or unexpected stress conditions, particularly in high cell density fed-batch and continuous cultures.

The prediction capacities of the models for these cultures are generally poor. To combat this situation, more quantitative knowledge on intracellular processes and structured models based on in vivo fluxes and intracellular parameters are needed. In particular, more research is needed to understand and to mathematically describe the kinetics and dynamics of the cell death machinery. It has been generally demonstrated that most commercially important mammalian cell lines primarily die by apoptosis in the physical and chemical environments of bioreactors. In this respect, the rapid development of cell and molecular biology concerning the genetics, biochemistry, and physiology of apoptosis gives a good chance to develop more comprehensive structured kinetic model for the growth and death of cell culture.

Till now, the intrinsic kinetic relationships among growth (cell cycle), death, product formation, and quality are not well understood. In view of the importance of glycoproteins as pharmaceutical products, the kinetics of glycosylation of proteins deserves more attention. Due to the diversity of glycosylation patterns and the many factors involved in the post-translational modifications this represents a challenge for kinetic modeling.

Another challenge for model development is the incorporation of genomic and proteomic data into kinetic analysis. Presently, there is a flood of genomic and proteomic data produced for many different cells. However, they normally only give static or structural information on the genes and proteins possibly involved in the cellular processes. The mechanisms and kinetic control of the functionality of the many genes and proteins under cell culture conditions remain to be explored. Research into these aspects is not only important for a more predictable and targeted improvement and control of animal cell culture but will have impact in many other areas of life science. In fact, understanding genome-wide kinetic behavior of cells is one of the central themes of postgenomic research. Cell culture as a clearly defined and reproducible biological system may be a good model system for this endeavor.

5

Stem Technique in Mice

The gene trap approach is based on the use of mouse *embryonic stem* (ES) cells and a class of vectors containing a splice-acceptor site upstream of the reporter and resistance genes, β-galactosidase (β-gal, *lac*Z) and neomycin resistance (*neo*), respectively. Integration of these vectors into a genomic locus, controlled by a functional promoter, results in the generation of a fusion transcript between the endogenous gene and the *lac*Z gene. Fusion mRNAs transcribed from the promoter of the tagged locus mimic endogenous gene expression, which can be monitored by visualizing *lac*Z activity. The tagged genes can be identified by the use of anchored *polymerase chain reaction* (PCR) procedure, performed on RNA extracted from the selected ES cell clones. In addition, the gene trap vectors could also act as insertional mutagens.

After genetic manipulation by electroporation or retroviral infection, the cells remain pluripotent, and when reintroduced into morula by embryo-aggregation (or into blastocyst by microinjection), they retain the potential to contribute to all tissues of the mouse including the germ line. Using this strategy, the expression patterns of the endogenous genes during embryogenesis and adulthood, as well as the phenotypic consequences of the gene trap mutations, may be analyzed.

One major flaw, of the gene trap approach we have used, is the necessity to generate a large number of mice from the corresponding ES cell clones to obtain few interesting genes. One way to circumvent this problem is offered by the possibility to preselect gene-trapped ES cell lines in vitro before generating the mice. This could be extremely useful when searching for specific classes of genes, for example, genes activated by specific signaling cascades. A simple and reproducible preselection protocol has been set up in our laboratory, in which gene-trapped ES cell lines are screened for the β-gal staining patterns in vitro and tested for their responsiveness to specific growth–differentiation factors such as follistatin, nerve growth factor, and retinoic acid.

Huge effort is being carried out worldwide to create gene trap libraries, which could, in principle, saturate the mammalian genome. We believe that the gene trap approach will increase our ability to study gene activity in depth and to associate such genes to human diseases.

In this chapter, we describe the "*routine*" gene trap methodology which has allowed us in the last years to identify, clone, and mutate several genes important for vertebrate development and cell biology. We will not discuss here the morula aggregation technique, extensively described elsewhere, but we will focus on ES cells manipulation and on the gene expression screening which follows the generation of a chimaeric mouse and precedes the phenotypic analysis of the mutant allele.

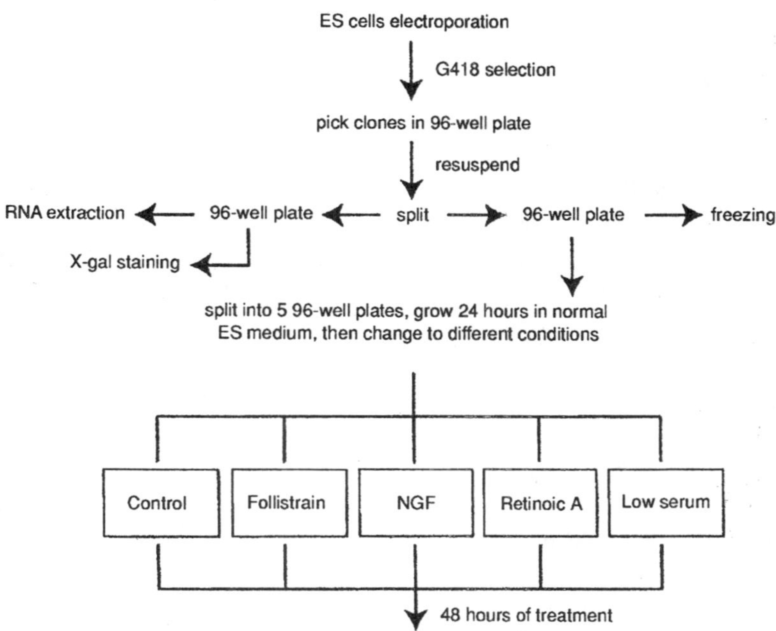

Fig. 5.1. ES clones in vitro preselection. Diagram of the screening used in our laboratory to preselect gene-trapped ES cell lines with soluble factors.

Materials

1. ES cell medium: Dulbecco's modified Eagle medium (DMEM), 100 μM β-mercaptoethanol, 2 mM glutamine, 1% stock solution of nonessential aminoacids, 1 mM Na-pyruvate, 15% (v/v) *fetal calf serum* (FCS), 500 U/mL *leukemia inhibitory factor* (LIF).
2. Phosphate-buffered saline (PBS) filter-sterilized. Composition for 1 L: add distilled water to 8.0 g NaCl, 0.2 g KCl, 1.15 g $Na_2HPO_4 \bullet 2H_2O$ and 0.2 g KH_2PO_4, adjust pH to 7.2, and add distilled water to 1 L.
3. Mitomycine C. Stock concentration is 2 mg/vial diluted in 2 mL 1X PBS, filter-sterilized, and stored at 4°C in the dark. The inactivation medium is made of 10 mL DMEM, 10% FCS, 100 μL mytomicine C stock solution.
4. Gene Pulser II electroporation apparatus.
5. 1X Trypsin-EDTA.
6. Geneticin (G418): 200 μg/mL, dissolved in dH_2O.
7. ES cells fixing solution: 0.2% glutaraldehyde, 2 mM $MgCl_2$, 100 mM K_2HPO_4 (pH 7.4), 5 mM EGTA.
8. ES cells washing solution: 0.01% Na-deoxycholate, 0.02% Igepal CA-630, 2 mM $MgCl_2$, 0.1 M K_2HPO_4 (pH 7.4), 5 mM EGTA.
9. 5-Bromo-4-chloro-3-indolyl β-D-galactopyranoside (X-gal).
10. ES cells staining solution: 0.5 mg/mL X-gal, 10 mM $K_3[Fe(CN)_6]$, 10 mM $K_4[Fe(CN)_6]$, 0.01% Na-deoxycholate, 0.02% Igepal CA-630, 5 mM EGTA, 2 mM $MgCl_2$, 0.1 M K_2HPO_4 (pH 7.4).
11. 2X ES cells freezing medium: for 10 mL, 1 mL dimethyl sulfoxide (DMSO) cell culture grade, and 2 mL FCS to 7 mL ES cells medium.
12. Dissecting microscope.
13. All *trans* retinoic acid (retinoic A).
14. Follistatin.
15. Nerve growth factor-β.
16. TRIzol.
17. 1:1:24 Phenol:Chloroform:Isoamylalcohol.
18. Chloroform.
19. Isopropylalcohol.
20. 70% Ethanol.
21. Lauryl sulfate sodium salt (SDS).

22. Ethanol absolute.
23. 5' RACE System for Rapid Amplification of cDNA Ends kit, Version 2.0.
24. 0.1-μm nylon filters.
25. pGEM-T Vector System.
26. Tris-EDTA buffer (TE): 10 m*M* Tris-HCl, pH 8.0, 1 m*M* EDTA, pH 8.0.
27. 8 m*M* Ammonium-acetate.
28. Electrocompetent cells ElectroMAX DH5α-E.
29. Transfer membranes Biodyne A.
30. N, N-Dimethylformamide.
31. Formaldehyde.
32. Glutaraldehyde.
33. X-gal staining solution for tissues (embryos and adult brains):

	10 mL	25 mL	50 mL
1X PBS	9.05	22.625	45.25
X-gal (40 mg/mL in N,N-Dimethylformamide)	0.25	0.625	1.25
$K_3[Fe(CN)_6]$ (200 m*M*)	0.25	0.625	1.25
$K_4[Fe(CN)_6]$ (200 m*M*)	0.25	0.625	1.25
$MgCl_2$ (100 m*M*)	0.2	0.5	1.0

34. Fixative solution for tissues (embryos and adult brains):

Fixative A (FixA):	25 mL	50 mL	100 mL
37% Formaldehyde	0.675	1.35	2.7
25% Glutaraldehyde	0.2	0.4	0.8
10% Igepal CA-630	0.05	0.1	0.2
20X PBS	1.25	2.5	5.0
dH_2O	22.825	45.65	91.3

Fixative B (FixB):	25 mL	50 mL	100 mL
37% Formaldehyde	0.675	1.35	2.7
25% Glutaraldehyde	0.2	0.4	0.8
10% Igepal CA-630	0.5	1.0	2.0
1% Na-deoxycholate	2.5	5.0	10.0
20X PBS	1.25	2.5	5.0
dH_2O	19.875	39.75	79.5

35. Glycerol 15, 30, 50, 70, and 80% in PBS (v/v).

METHODS

Electroporation of ES Cells

1. Wash the confluent ES cells with PBS.
2. Add 1X Trypsin-EDTA (2 mL for an 8.5-cm culture dish) and incubate for 5 min at 37°C.
3. Pipet the cells up and down in order to obtain a single-cell suspension.
4. Add an excess of ES cell medium (6 mL for an 8.5-cm culture dish) to stop trypsinization, and mix well.
5. Spin down the cells with a table-top centrifuge at 260*g* for 5 min.
6. Remove the supernatant and resuspend the cells in new ES cell medium.
7. Split the cells on newly inactivated *mouse embryonic fibroblasts* (MEF) feeder plates.
8. After 24 h, change ES cell medium.
9. Twenty-four hours after changing the medium, trypsinize the cells as described above and add ES cell medium to 10 mL, mix well, and spin down the cells as above.
10. Aspirate the supernatant and resuspend the cells in 30 mL 1X PBS. Determine the cell number. The optimal number is 1.5×10^7 cells.
11. Spin down the cells as in step 5 above and resuspend them in 0.8 mL 1X PBS containing 25 μg/mL of linear DNA. Incubate the suspension at room temperature for 5 min.
12. Pipet the cells gently several times up and down and transfer 0.8 mL of the suspension to one electroporation cuvette, carefully avoiding air bubbles. Electroporate with one pulse of 500 μF and 250 V at room temperature. Incubate at room temperature for 5 min.
13. Transfer the cells to ES medium (to a final vol of 30 mL) and plate them on 7 × 8.5 cm newly inactivated MEF feeder plates. As a control plate, use cells at the same density electroporated without linear DNA.
14. After 24 h, change ES cell medium to a medium containing G418 for positive selection.
15. Change medium every day for 9 d.
16. Ten days after electroporation, check the control plate, which should not contain any ES cell clones. Large ES cell clones (>1000 cells) can be picked.

Picking of ES Cell Clones

1. Mark a circle under the clones to be picked.
2. Prepare a 96-well plate with 35 μL 1X trypsin-EDTA in each well.
3. Under a sterile hood and using a dissecting microscope pick, with a P20 Gilson pipet equipped with 10-μL PCR tips, every marked clone individually and transfer to the 96-well plate. Incubate at room temperature.
4. Check trypsinization under the microscope every 2 min, gently shaking the plate. When the clones are disaggregated into small cell clumps, add 65 μL ES medium and pipet several times up and down in order to obtain a single-cell suspension.
5. Transfer 50 μL of each suspension, using a multichannel pipet to two other 96-well plates with newly inactivated MEF feeders in 100 μL ES medium. One will be used for preselection experiments and for freezing the clones, the second for X-gal staining and RNA extraction.
6. First plate. After 3 d, the clones will be ready for freezing and in vitro preselection. After trypsinizing the cells, from the first 96-well plate, with 35 μL 1X trypsin-EDTA for 5 min, add 2X freezing medium to 25 μL of the cell suspension, mix well, wrap the plate with paper, and store at –80°C in a styrofoam box. The cells will remain viable for several months.
7. Add 90 μL of ES medium to the remaining 10 μL of suspension.
8. Plate the suspension on a 96-well plate containing newly inactivated MEF feeders in 100 μL ES medium for pre-selection experiments.
9. After 3 d of incubation, expand the cells into five different 96-well dishes without MEF feeders, and culture with normal ES cell medium.

Screening of Gene Trap ES Cell Lines Responsive to Soluble Factors

1. Let the 96-well plates as described above, step 9 grow for 24 h. When small colonies are apparent, medium is removed, and the dishes are cultured under different conditions.
2. Add ES medium without LIF according to the following five different conditions, successfully applied in our laboratory:
   (a) ES medium with 20% FCS (control plate).
   (b) ES medium with 20% FCS and 150 ng/mL follistatin (follistatin plate).

   (c) ES medium with 1% FCS (low serum control plate).
   (d) ES medium with 1% FCS and 100 ng/mL NGF (NGF plate).
   (e) ES medium with 1% FCS and 0.2 m*M* retinoic A.
3. Culture cells for additional 48 h.
4. Process for X-gal staining according to as described below.

X-Gal Staining of ES Cells Clones

1. Second Plate. After 3 d, the clones will be ready for X-gal staining. Trypsinize the cells from the second 96-well plate as describe above, step 5, with 35 μL 1X Trypsin-EDTA for 5 min.
2. When the clones are disaggregated into small cell clumps, add 65 μL ES medium, and pipet several times up and down in order to obtain a single-cell suspension.
3. Transfer 50 μL of each suspension, using a multichannel pipet, to another two 96-well plates with newly inactivated feeders in 100 μL ES medium.
4. After three d, wash the cells from one plate in 1X PBS.
5. Fix the cells up to 3 min at room temperature in fixing solution.
6. Wash 3 to 4 times in washing solution at room temperature.
7. Incubate in staining solution at 37°C for up to 3 d. Stop the reaction by washing in 1X PBS.
8. Refix for 5 min in 2% formaldehyde in 1X PBS.
9. Wash 3 to 4 times in 1X PBS and store at 4°C.

Identification of the Tagged Gene by the 5' RACE Procedure

RNA extraction from ES cells

1. After 3d, trypsinize the cells from the second 96-well plate as described above, with 35 μL 1X trypsin/EDTA for 5 min.
2. When the clones are disaggregated into small cell clumps, add 65 μL ES medium and pipet several times up and down in order to obtain a single-cell suspension.
3. Transfer the suspension with a multichannel pipet to a 24-well plate with newly inactivated feeders in ES medium.
4. After an additional 3 d, expand the cells on a 6-well plate and let them grow to confluence.
5. Wash cells twice with 1X PBS.
6. Add 0.75 mL of Trizol to a single well, swirl the plate for at about 30 sec, and collect the lysate in a 1.5-mL tube, passing it several times through a pipette.

7. Incubate the samples for 5 min at room temperature.
8. Add 0.14 mL of chloroform, shake the tube vigorously for 15 s, and incubate the sample for 2 to 3 min at room temperature. Centrifuge the sample at 12,000*g* for 15 min at 4°C. Following centrifugation, take the upper aqueous phase and transfer to a new tube.
9. Precipitate the RNA by mixing with isopropylalcohol (0.375 mL). Incubate the sample for 10 min at room temperature and centrifuge at 12,000*g* at 4°C for 10 min. The RNA precipitate will form a gel-like pellet.
10. Discard the supernatant and wash the pellet once with 0.75 mL of 70% ethanol.
11. At the end of the procedure, air-dry the pellet, keeping the lower half of the tube in dry ice. Do not use a vacuum. Dissolve the RNA in 20 μL RNase-free water, 0.5% SDS solution by repeated pipetting and incubating for 5 min at 55°C with gentle shaking.

5' RACE procedure

We have used the protocol provided from the Gibco Instruction manual 5' Race System for rapid amplification of cDNA ends, Version 2.0, derived from Frohman et al., with some modifications:

1. Superscript II RT reaction. Problems with 5' RACE due to the secondary structure of the target RNA can be easily avoided by increasing the temperature of the *reverse transcription* (RT) reaction. The gene trap approach is aimed to identify novel mRNAs whose structural features are unknown to the investigators. Therefore, during the first strand synthesis step, we shifted the primer/RNA mix directly from 70°C to 50°C and prewarmed the complete reaction mixture to 50°C, before adding it to the primer and RNA.
2. cDNAs size selection. To size-select cDNAs, in order to amplify only the most sequence-informative fragments, the reactions were spotted after dC-tailing on 0.1-μm nylon filters. Microdialysis took place against TE for 3 h, allowing the discharge of DNA fragments less than 200 bp. Size-selected cDNAs were then taken through two rounds of nested PCR amplification and microdialyzed after each round.
3. Isolation of 5' RACE products. Extract the PCR samples twice with phenol:chloroform:isoamylalcohol, once with chloroform, and precipitate the RACE products on ice for 10 min by adding 50 μL

TE, 33 μL of 8 *M* ammonium-acetate, and 330 μL of ethanol. Centrifuge for 10 min at room temperature at 12,000*g*, and wash the pellet with 70% ethanol. Dry the pellet and resuspend in 20 μL TE.

4. Cloning of 5' RACE products. We have used the pGEM-T Vector system from Promega, based on the concept of T/A overhang. Set up the ligation reaction as described below:
5. Use 0.5-mL tubes with low DNA-binding capacity.

T4 DNA ligase 10X buffer	1 μL
PGEM-T Vector	1 μL
PCR product	2 μL
T4 DNA Ligase (1 U/μL)	1 μL
dH_2O to a final volume of	10 μL

6. Incubate overnight at 15°C.
7. Remove a 2-μL aliquot for the transformation step.
8. Transform into high efficiency competent cells.
9. Screen colonies for the positive inserts by colony hybridization with a labeled internal oligonucleotide.
10. Sequence >20 clones and confirm the reliability of the clones by comparing with the most common databases and excluding clones with spurious sites of fusions.

Production of Mouse ES Cells Chimaeras by Morula Aggregation

The generation of ES cell-derived chimaeric mice by blastocyst injection or morula aggregation is not described in this chapter. Upon transfer to a pseudopregnant recipient, the ES cells participate in normal development of the chimaeric embryo and contribute to all cell types, including the germ line. Once germ line chimaeras are generated, they represent sources for heterozygous mice carrying the inserted trapping vector. The activity of β-gal can be detected in whole embryos and in sectioned *postnatal* (pn) tissues by a chemical reaction that involves the cleavage of the X-gal substrate by the β-gal (*lac*Z product) encoded by the transgene. This reaction leads to a blue coloring of positive tissues, as described in the next section.

Screening for lacZ Transgene Expression during Embryogenesis

1. Preparation: embryos or tissues are dissected in cold 1X PBS (4°C).
2. Embryos are fixed according to Table 5.1.
3. Postnatal brains (7–10 pn) are fixed for 30 min in FixB, embedded in paraffin wax, cut in 100-μm sections, fixed for an additional 1

h in fresh FixB. Older brains (more than 3 mo) are prefixed for 45 min, and then sectioned and refixed in fresh FixB for additional 60 min. Fixation should occur always on ice with gentle shaking, and the fixative should be freshly made.

4. Washing and X-gal staining: The embryos or tissues are washed twice for 20 min in 1X PBS at room temperature.
5. Add the staining solution. The staining is performed overnight in the dark at 30°C.
6. Clearing: The staining is stopped with two washes of 1X PBS. The embryos are then cleared by subsequent washes in 15, 30, 50, 70, and 80% glycerol-1X PBS. Each washing step should be done for 1 d.
7. Staining patterns are then classified for both spatial and temporal differences between mouse lines, with particular attention to the domain of interest of the laboratory. Because this method utilizes an extremely simple enzymatic reaction, the full characterization of the expression pattern of the captured gene can be completed in an efficient manner.

Table 5.1. Fixation of embryos from different developmental stages

Embryonic age	*Fixation time*	*Fixative solution*
Day 9.5 p.c.	30 min	FixA
Day 10.5 p.c.	30 min	FixA
Day 11.5 p.c.	50 min	FixA
Day 12.5 p.c.	50 min	FixA
Day 13.5 p.c.	30 min	FixA; cut in the middle
	30 min	Fresh FixA
Day 15.5 p.c.	30 min	FixA; cut in the middle
50 min	Fresh	FixA

Notes

1. To prepare the MEF feeder, proceed as follows:
   (a) Remove about ten embryos at d 13 *postcoitum* (p.c.). Remove and discard the head, the liver, and the internal organs. Remove as much blood as possible by washing the bodies twice in 1X PBS.
   (b) Cut the bodies into small pieces, and wash over a sieve with 1X PBS.

(c) Incubate in 50 mL 1X trypsin-EDTA solution in an erlenmeyer flask containing glass beads for 30 min at 37°C under gentle agitation on a magnetic stirrer.

(d) Add other 50 mL of 1X trypsin-EDTA solution and let shake for additional 30 min.

(e) Pour the MEF suspension over a sieve, collect in 50-mL Falcon tubes, and centrifuge for 10 min at 260*g*.

(f) Plate 5 × 10^6 cells/150-mm plate. After 24 h, change the medium. When semicon- fluent, freeze the cells (1 plate/1 vial).

(g) For inactivation, revive 1 vial of MEFs in a 100-mm plate.

(h) After about 48 h, split to 5 × 100-mm plates.

(i) After about 48 h, split each 100-mm plate to a 150-mm plate.

(j) When confluent, treat with mitomycin C. Remove the medium and add 10 mL inactivation medium to each 150-mm plate. Incubate about 2.5 h, change the medium 3 times. After 24 h, inactivated MEFs should be trypsinized and plated as follows: 60-mm plates, 2 × 10^6 cells; 100-mm plates, 4.5 × 10^6 cells.

(k) When the MEFs are adherent (after about 5 h), the ES cells could be plated over the feeder.

2. The rationale behind the use of reporter constructs is to tag and detect *cis*-regulatory sequences by locating the reporter gene within an endogenous gene. The reporter construct may also disrupt endogenous gene function and act as a mutagen. All trapping vectors devised so far consist of modifications on one basic structure. Endogenous genomic sequences are placed next to a reporter gene, which is then brought under their influence. In those vectors, the 5' region contains a splice-acceptor site derived from the *engrailed2* (*eng2*) gene (*eng2* intron element [ei] and exon element [ee], respectively). The reporter gene is *lac*Z, and the selectable marker is neomycin phosphotransferase (*neo*) fused together into the *lac*Z/*neo* gene (βgeo). The presence of an *internal ribosomal entry site* (IRES) derived from the encephalomyocarditis virus, allows the reporter and resistance gene to be expressed without requiring an in-frame fusion with the coding region of the endogenous gene. Salminen et al. have constructed a new poly(A) trap vector. In IRESβgalNeo(-pA), the transcription of *neo* is under the control of the β-*actin* constitutive promoter (βa), while the β-gal expression is dependent on the activity of the trapped gene. A *Pax2* splice-donor element is placed at the 3' of the bicistronic structure.

When the ES cells are cultured under conditions that allow differentiation, a 3-fold increase in positive-stained clones is observed with IRESβgalNeo(-pA). As a consequence, this vector will be very useful in the search for genes responding to specific regulatory factors in vitro.

3. Cells should not be too confluent at time of freezing. Freeze gradually: 2 h at –20°C, then store at –80°C. To resuspend, thaw quickly in a waterbath at 37°C, add 100 μL of medium, and let the cells grow.
4. Staining was carefully compared for each cell line in the 5 different dishes to detect gene trap lines that responded to one (or more) factor(s). Selected cell lines were then thawed, expanded to 35-mm plates, and screened again with the above protocol to confirm their response to the factors.
5. In order to optimize the quality of the RNA template used for RT reaction, we performed 16 different RNA preparations using four different cell lines and four different RNA preparation protocols. The ES cell lines were divided onto four groups taking into account the different degrees of relative abundance of the trapped mRNAs, as revealed by X-gal staining of cell colonies; four lines were taken as samples, one from each group. Three commercial kits for RNA preparation were used. We report here that the best quality total RNA from ES cell lines grown to confluence in 6-well plates, measured by Northern blot and PCR, was obtained using 0.75 mL Trizol. As reported by Townley et.al. an additional phenol extraction, prior to RNA precipitation, was found to improve the quality of the RNA.
6. First strand cDNA. The reaction was carried out with random primers and with vectorspecific primers. Although PCR performed using the equally abundant hypoxanthine phosphoribosyltransferase (HPRT) mRNA specific primers has revealed that a certain level of unspecificity is generated also by the vector specific primers, this is the most useful approach.
7. Southern blot analysis of 5' RACE products. In order to make a strategic choice between the direct sequencing of 5' RACE products and their cloning prior to sequencing, we have tested the degree of purity of the amplification products obtained with this method. Five microliters from the second round RACE products were run on 2% agarose gel and transferred to transfer membranes. Following the transfer, the filter was probed with a labeled 42-

mer primer corresponding to the *eng2* splice-acceptor site, present in all the analyzed vectors immediately 3' of the integration site. Hybridizations were performed at 65°C. Filters were washed in stringent conditions and exposed to film for 2 h. Although positive hybridization signals were visible in all lanes of the electrophoresis gel, they either did not correspond to the more represented bands or there was more than one signal in one lane. If it can be argued that multiple bands represent different sized amplification products of the same cDNA, this is uncertain *a priori* and does not ensure a safe way to routinely proceed. In only a very few cases, we observed a sharp band that represented a reasonably good template for direct sequencing.

8. It should be mentioned that the expression of the reporter gene in some cases does not mimic that of the endogenous trapped gene. This could be caused, for example, by the disruption of intronic regulative elements of the endogenous gene by the gene trap insertion. It is, therefore, essential to confirm the expression of the trapped gene, once it has been successfully cloned, by *in situ* hybridization.

Functional Genomics by Gene-Trapping in Embryonic Stem Cells

The pace of sequencing whole genomes by far exceeds our increase of knowledge on gene functions. Twenty-two genomes have been already completed and both the human and mouse genome are close to be completed as well. The number of sequences in the *National Center for Biotechnology Information* (NCBI) gene bank doubles approximately every other 15 months. The gain of information on *expressed sequence tags* (ESTs) from a multiplicity of organisms and cell types rose exponentially and is close to reaching saturation.

Almost two million human and one million mouse ESTs can be found inside the NCBI database by today. In fact, over 63% of the 4.6 million different sequences in the Genbank are ESTs.

In the post-sequencing area, large-scale and high-throughput technologies have to be developed and applied in order to understand the physiological role of any given gene–protein in a living organism. Computational analysis based on the identification of homologous genes in different species will be necessary, but not sufficient. In reported cases, homologous genes fulfill completely different functions in different species.

An impressive demonstration of this phenomenon has been given by the functional comparison of similar molecules in the two closely related nematodes *P. pacificus* and *Caenorhabditis elegans*, which revealed that homologous genes can be recruited to serve completely new functions in new regulatory linkages. While remaining in the original developmental program, genes were found to have changed their molecular specificity and at the same time, retained other functions. Genes have jumped in or moved out of regulatory circuits and are suddenly expressed in a spatially and temporarily different manner. Therefore, mere sequence or structural similarity does not necessarily allow conclusions on possible protein functions.

It is clear that the actual function of each gene needs to be studied in the entire organism. The most straightforward way has been to analyze the phenotype of organisms carrying either gain- or loss-of-function mutations within the gene of interest. Traditionally, the phenotype-driven approach (forward genetics) has been most fruitful in organisms with comparably small genomes.

The source of natural mouse mutants is relatively limited. At present however, the resource of natural mutants experiences a remarkable increase by large-scale chemical mutagenesis screens with subsequent screening for dominant and recessive mutants. *Ethylnitrosourea* (ENU)-screens are perfect tools in order to identify causes of monogenic disease. Redundant factors however, as well as complex traits, are falling through its net. The most important disadvantage in the analysis of naturally and chemically produced mutants is, however, that the identity of the responsible gene has to be subsequently determined. Although improvements have been made, the identification of point mutations can still turn out to be a tedious venture and can take several years in individual cases.

The targeted gene inactivation, which is now being performed, since more than 12 yr in the mouse has led to an exponential growth of information on the function of a multiplicity of genes and has established the mouse as the favorite organism when considering the generation of models for human genetic disease. The number of the published mouse genes with a targeted mutation, so far, can only be estimated according to the current mouse locus catalogue around two thousand. The biomed knock-out database shows 1818 entries for the term "*knock-out.*"

A satisfactory answer to the question on gene function however, can often not be given from one single knock-out animal. In general, only the earliest phenotypes will be determined by conventional gene inactivation experiments and will, in many cases, lead to lethality. Depending upon construct design, different mutations have led to variable phenotypes and it turns out that, increasingly, a promoterdriven resistance cassette is often disturbing to neighboring genes, which makes an interpretation of the mutant phenotypes even more difficult. Therefore, it appears that a number of knock-outs that have been performed in the past need to be revisited in a complementary approach.

It is important to point out, however, that despite world-wide activities in the past 10 yr, no more than 5% of the assumed mouse genes have been inactivated, which would add up to 200 yr or so in order to study each gene by applying conventional knock-out technology. The knock-out technology has experienced substantial improvements from different areas. Particularly, the construction of knock-out vectors could be shortened from several months initially, to only a few weeks at present.

The possibility of a high through-put functional analysis, which is capable of complementing the current genome sequencing, can be accomplished by a large-scale insertional mutagenesis, namely the gene-trapping. Therefore, in the following chapter we want to give a general overview on the latest development of gene-trapping activities in mouse *embryonic stem* (ES) cells, highlighting in particular, the different approaches, advantages, and accomplishments, as well its inherent difficulties and pitfalls.

Gene-Trapping Strategies

Enhancer trap

The trapping principle is based on random insertional mutagenesis of a vector ideally containing a promoterless reporter gene. The trap vector both serves as a sequence tag for the identification of the mutated locus and can provide a reporter–selection marker for monitoring endogenous gene activity. Originally, by utilizing bacteriophage transposable elements, a reporter cassette provided with a minimal promoter has been introduced into the *Escherichia coli* genome. This enhancer trapping approach has been demonstrated to be straightforward, for both the identification and mutation of bacterial genes. The original enhancer trap approach, which had been successfully applied to the bacterial genome, turned out to be less straightforward in more complex genomes. Although useful in order to identify tissue or cell type-specific regulatory elements, eucaryotic enhancers are often located far away

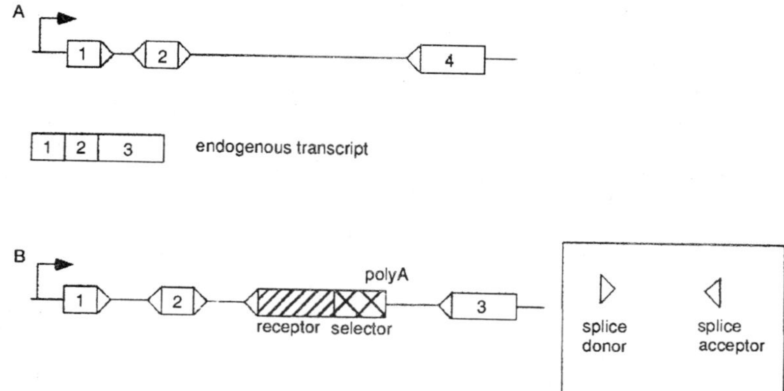

Fig. 5.2. The gene-trapping principle is based on random integration of a receptor-selector cassette into a genome to both mutate and identify the trapped locus. (A) Wild-type locus; (B) trap vector integration.

from the regulated genes. In addition, enhancer trap vectors are in general not mutagenic in eukaryotes.

Exon trap

In principle, exon trap vectors are without *splicing acceptors* (SA). Consisting only of a selector or reporter–selector cassette, their activation requires either in-frame exon integration or integration in 5' *untranslated regions* (UTRs) of actively expressed genes. Therefore, this vector type appears extremely inefficient in respect to identifying resistant colonies. Due to the limitations of exon trap vectors, mostly gene-trap vectors are being used for large-scale insertional mutagenesis screens in eucaryotic genomes.

Gene-trap

Typical gene-trap vectors are promoterless and function by generating a fusion transcript with the endogenous gene. The presence of *splice donor* (SD) or SA elements leads to the generation of a fusion protein, even in case of an intron integration.

Gene-trapping has been applied with great success in *Drosophila*, *C. elegans*, and murine cells. In *Drosophila*, insertional mutagenesis has been performed using a nonretroviral transposon called *P-element*, which can be utilized as an enhancer detector and sequence tag in order to both identify and isolate affected loci.

The development of the mouse takes place *in utero* and is not accessible directly to genetic manipulation. Since the breakthrough of the ES cell system, it suddenly became feasable to manipulate mouse embryonic cells ex vivo and generate animals carrying a single predetermined mutation. By combining both, insertional mutagenesis and the ES cell system, Gossler et al. have pioneered the gene-trapping approach in mammalian cells. Friedrich and Soriano subsequently generated twenty-four mouse strains carrying gene-trap mutations demonstrating the mutagenicity of the integrated vector in living organisms.

In higher organisms, gene-trapping turned out to be not only a tool for mutagenesis, but also as a means for gene expression studies and especially for gene discovery by capturing open reading frames. Besides ES cells, the gene-trapping principle has also been successfully applied to other cell types.

SA-type vectors

Most commonly used vectors contain a SA or both an SA and an SD. The typical SA type vector uses an ATG-less reporter–selector cassette, which is flanked by a 5' SA sequence. In eucaryotic cells, nuclear pre-mRNA introns are excised by a large ribonucleoprotein complex known as the splicosome, which recognizes sites at the 5' and 3' ends of the intron (the donor and acceptor splice sites, respectively). A minimal SA consists of a branch point sequence followed by a polypyrimidine tract and the actual splice site, which is recognized by the splicosome. Widely used SA in gene-trap vectors are the murine *engrailed-2* and the adenovirus late major transcript SA. More recently, the human *BCL-2* gene intron2/exon3 SA has been introduced. Proper integration and splicing will generate a fusion protein with the trapped locus. The fusion transcript will ideally be prematurely terminated at the poly(A) site of the selection–reporter cassette.

In order to identify genes independent of reading frames, an *internal ribosomal entry site* (IRES) from the encephalomyocarditis virus is utilized 3' of the SA.

SD-type vectors: Poly(A) trap

A limitation of the commonly used SA vector type is that mostly expressed genes can be identified. In contrast, the poly(A) trap vector type allows the identification of all intron-containing genes independent of their expression status. The simplest poly(A) trap vector contains a promoter-driven selector gene without a poly(A) sequence. Thus, drug resistance is only obtained after successfully capturing a poly(A) signal in the genome. Furthermore, poly(A) trap vectors have been developed that contain an additional SD sequence.

The latest poly(A) trap vectors contain, in addition, a reporter gene with a SA in front, followed by a selector gene driven by a strong promoter fused to a SD. SD vectors, thereby, allow to monitor the regulation of the trapped gene in vivo assessing reporter gene expression.

The advantage of SD-type vectors is their independence of actively expressed genes due to the ubiquitously active promoter of the selection unit. The only known genes that are not polyadenylated in metazoan organisms, are the major histone genes. Therefore, poly(A) trap vectors will basically allow the entrapment of most genes. A variety of SD sequences that are derived from the mouse *Pax-2* and Hprt genes, and a synthetic SD consensus sequence, have been successfully utilized.

Reporter–selector cassettes

Several reporter assays have been described that are applicable for high-throughput screening procedures.

A widely used reporter gene for high-throughput screening is the bacterial β-galactosidase (β-gal) (lacZ), which has the advantage to be well characterized, stable, nontoxic, and additionally, has inexpensive substrates. Highly sensitive fluorescent and chemiluminescent substrates are available for β-gal. A limitation of β-gal is some endogenous activity in the mouse, starting around embryonic d 11.5. In older embryos or newborns, this problem may be circumvented by utilizing commercially available lacZ antibodies.

An alternative to β-gal is the bacterial β-lactamase, for which no endogenous activity has been described in the mouse. Human placental alkaline phosphatase (AP) has been utilized successfully in gene-trap vectors.

The jellyfish *green fluorescent protein* (GFP), which is available in multiple forms, has the advantage that no substrate is needed and no endogenous activity is observed in murine cells. In addition, antibodies for GFP detection postmortem are commercially available. Recently though, evidence of GFP toxicity in mouse cells has been reported when highly expressed.

Resistance markers in gene-trap vectors are traditionally neomycin phosphotransferase or lacZ/neo fusions (βgeo), but other resistance markers like phleomycin have been used successfully. LacZ/hygromycin cassettes have been utilized in order to trap genes in transfected cells and cells carrying a targeted mutation. Poly(A) trap vectors are available with a neomycin or puromycin resistence cassette.

Bias of vector integration

Apparently, the integration of a trap vector is not entirely random. Often, integrations into the 5' end of genes have been observed. 5' integrations result in relatively small fusion proteins, whereas 3' insertions can lead to large fusion complexes between the resistance marker and the trapped gene. Therefore, in case of 3' integrations, because of inadaequate protein folding, the resistence marker may become inactive. Accordingly, SA trap vector integrations could be more or less random, but 3' insertions might simply not be functional. Given, that this hypothesis is correct, one would expect 3' integrations to be functional for SD vectors. In contrast however, for SD-type vectors, it has been reported for 105 traps in total that 88% of integrations occured either 5' or in the middle of the encoded cDNA.

Chowdhury et al. did not observe a preferred 5' integration when using the SA-type vector types SAβgeo and IRESβgeo. Instead, integrations occured unbiased among 55 examined cDNAs. In the report however, the authors did not distinguish between vector integrations with and without an IRES sequence. Interestingly, we have found all 3' integrations described within the report of Chowdhury et al. to be derived from the IRES vector. Misfolding reasons, though, should not apply to IRES vectors, since translation of the selector gene takes place independently of the trapped locus from a bicistronic transcript.

Bronchain et al. have addressed the misfolding problem by introducing a glycine stretch 5' of a GFP reporter cassette for a gene-trap approach in *Xenopus*. It has not been reported yet, whether the glycine-stretch has an impact on the 5'–3' integration ratio.

Only the analysis of larger numbers of gene-trap integrations will give a satisfactory answer to this issue. Assuming that there is a preference to identify 5' integrations, this will have the positive side-effect of generating preferably null mutations, since 3' integrations might only partially disrupt the function of the trapped gene.

Advantages and Disadvantages of Individual Vector Types

SA-type vectors

Each gene-trap vector has its own inherent advantages and disadvantages. A limitation of both the SA-type as well as the exon trap vectors is mainly the dependency on integration in active genes. It has been speculated that a large number of genes are expressed in ES cells probably due to an undermethylated status, and the analysis of 7000 gene-trap integrations in our database revealed that we are still far from saturation.

Trap vectors have been used without SA but with an IRES sequence instead. 5' rapid amplification of cDNA ends (RACE) sequences from gene-traps of this type usually yield an intronic sequence, which is under-represented in the public databases. IRES vectors without an SA which have integrated into intronic regions, have sometimes failed to be mutagenic due to splicing around the integration site or of unknown reasons.

SD-type vectors

Very little data is available on mouse strains carrying a poly(A) trap mutation. In contrast to the SA-type vectors, it has been speculated, that the integration of ubiquitously active transcriptional units could interfere with neighboring genes and thereby complicate the interpretation of phenotypic observations. Therefore, it is advised to finally remove the promoter element of the gene-trap vector by taking advantage of the Cre or Flp recombinase systems.

From the view of functional genomics, one of the major issues concerning the poly(A) trap approach is whether the sequences that are being discovered are actually coding for proteins. The integration in the vicinity of canonical and noncanonical poly(A) sites, which are not related to genes, appears as a hypothetical disadvantage of the poly(A) trap approach. This integration might lead to a high number of false positive gene-trap clones. The consensus sequence for a canonical poly(A) site is AATAAA, but variations of GU or U-rich sequences are productive and may also lead to proper polyadenylation. Recent studies of public EST databases have shown that the consensus AATAAA hexamer is absent from more than half of all 3' UTRs. Taking into account only the canonical site, the statistical occurence is 4^6, meaning that every other 4096 bp statistically, there is a poly(A) signal. This adds up to approximately 750,000 canonical poly(A) sites within the mouse genome. Given that two noncanonical sites are also capable of polyadenylating a transcript, this number adds up to more than two million sites. This number exceeds the estimated number of mouse genes by almost 20–30 times.

Naturally, there is more to proper polyadenylation than just a poly(A) signal. The poly(A) signal needs to be followed by a more complex signal that is not yet fully characterized. In particular, the RNA polymerase needs to be released from the DNA template, and the transcript needs to be cleaved to be polyadenylated. Therefore, the number of false positives will be significantly smaller than 30 times. But in a high-throughput screen, this fact may lead to an unacceptable number of false positive clones, which could then only be identified after sequencing by the fact that no splicing has had occured.

This limitation of poly(A) trapping has been recently addressed by the use of mRNA instability elements 3' of the SD, which lead to the degradation of polyadenylated transcripts, that have remained unspliced. It has been demonstrated that the use of instability elements diminishes the total number of clones by 50%.

Additionally, in the case of poly(A) trap vectors, single vector integrations are mandatory. Therefore, retroviral vectors should be used for vector delivery, since tandem integrations can lead to splicing into the neighboring reporter cassette and can, thereby, lead to the isolation of false positive clones.

Specialized trapping strategies

It has been estimated, that in humans, secretory molecules account for about onetenth of the human proteome. Those include major factors of signaling pathways, blood coagulation, immune defense and carcinogenesis, as well as digestive enzymes and components of the extracellular matrix and blood plasma. It has been speculated, that β-gal activity is lost when βgeo is fused to a signal peptide. Therefore, insertions in secreted molecules may not be detected. In order to overcome this constraint, a gene-trap vector of the SA-type provided with a transmembrane domain has been developed, which prevents the reporter–selector fusion protein from being secreted after integrating into a intercellular signaling molecule gene. Alternatively, the use of an IRES sequence will lead to an independent translation of βgeo from the bicistronic fusion transcript. Therefore, the reporter–selector protein should remain inside the cell.

The application of exogenous retinoic acid is known to induce profound effects on patterning of various tissues. In order to identify genes that are initially inactive in ES cells, but respond to challenges like retinoic acid, transcription factors, or signaling molecules in undifferentiated or differentiated ES cells, the induction trap strategy has been developed. In vectors used for an induction screen, the seletion marker is generally driven by a ubiquitously active promoter, whereas the reporter cassette is promoterless. As expected, only very few resistant clones (0.2–5%) are initially lacZ positive. Clones that turn "blue" in response to the challenge can be isolated and further characterized during this screen. In case a βgeo type of vector is used, only expression level difference can be monitored.

More recently, the induction strategy has been developed further to identify target genes of transcription factors. This approach is based on the internalization and nuclear addressing of exogenous homeodomain-containing proteins, using an *Engrailed homeodomain* (EnHD) as an example. An ES cell gene-trap library has been screened to identify targets of *Engrailed*. In this screen, 8 integrated gene-trap loci responded to EnHD. One of the genes identified was the *bullous pemphigoid antigen* 1 (BPAG1). By combining in vivo electroporation with organotypic cultures, it has been shown that a BPAG1 enhancer–promoter is differentially regulated by *Engrailed* in the embryonic spinal cord and mesencephalon. This strategy can, therefore, be used to identify and mutate homeoprotein targets. Because homeodomain third helices can internalize proteins, peptides, phosphopeptides, and antisense oligonucleotides, the strategy should also be applicable to other intracellular targets for characterizing genetic networks and dissecting different pathways involved in a large number of physiopathological states.

Gene Identification

RACE polymerase chain reaction

In both the SD and the SA trap vectors, the site of integration can be identified by 5' (SA trap) or 3' (SD trap) RACE and subsequent sequencing. A major advantage of using RACE *polymerase chain reaction* (PCR) to identify the integration site, lies in the possibility to use automated RACE procedures. RACE products can be sequenced without subsequent cloning in case a tagged primer is being used for the last amplification.

Trapping with poly(A) trap vectors is advantagous in respect to RACE procedures, since 3' RACE is more robust than the 5' RACE. This observation is probably based on the fact that tailing of the PCR product is not necessary because a poly(T) primer can be used in addition to a primer derived from the vector. A second disadvantage of performing 5' RACE is the under-representation of 5' sequences in EST databases. This limitation will be overcome, with the sequencing of the entire mouse genome. At present, we find approximately 30% of the obtained 5' RACE sequences to be unknown.

Plasmid rescue

The utilization of plasmid rescue to identify the trapped locus is possible in case the bacterial *origin of replication* (ori) and ampicillin resistance are retained in the gene-trap insertion. Araki et al. report for a total number of 109 clones that were generated by electroporation, the presence of the bacterial *amp* and *ori* in only 37% of the cases.

In contrast, presumably, resulting from homologous recombination between the *long terminal repeats* (LTRs), when using retroviral vectors, 70–80% of examined clones retain the bacterial sequences. Automation of the subsequent cloning step, which is involved in the protocol prior to sequencing, has not yet been achieved, therefore this method does not seem to be applicable in a high-throughput screen at present.

SupF complementation

A retroviral entrapment vector of the SA-type has been described, which facilitates the isolation of trapped loci by taking advantage of the bacterial supF tRNA gene. The protocol involves cloning of genomic DNA fragments into λ phages, plating on a nonpermissive bacterial strain, and screening with internal probes.

Construction of phage libraries

In some reported cases, the rapid identification of trapped loci by simple PCR or cloning-based techniques did not turn out to be straightforward. In these cases, the construction of a phage library from mutant cells or animals, and subsequent screening with vector sequences, has yielded the desired information.

Mutagenicity of Gene-Trap Vectors

Ideally, the integration of the gene-trap knocks the endogenous gene out or at least down. The phenotypes of mutant gene-trap mice can be phenocopies of null mutants, but can also be more variable, depending on the site of integration and the integrity of the vector cassette in the genome.

Two parts of the gene-trap vector mainly influence its mutagenicity in respect of interrupting the endogenous transcript: *(i)* the relative strength of the SA is critical; and *(ii)* effective termination of the fusion transcript at the introduced poly(A) signal is required.

A large number of different SA sequences have been used in gene-trap vectors, which have been described elsewhere. However, all vectors described show to some extent "*splicing-around*" the inserted vector. Thus, wild-type gene product can be made, and splicing-around leads to variable hypomorphic mutations as compared to null mutants. Hypomorphic phenotypes of gene-trap mice are sometimes informative as compared to complete knock-outs. As an example, Couldrey and co-workers have reported a gene-trap insertion into the histone 3.3A gene, which resulted in partial neonatal lethality in only 50% of the examined cases, stunted growth, neuromuscular deficits, and male subfertility. The wild-type transcript was found to be reduced by 4–7 times in homozygotes, as compared to wild-type littermates. However, there are also examples in which no phenotype has been obtained.

Recently, a vector system has been developed, which allows to test the relative strength of SA sites in vitro. The vector pSAT consists of a ubiquitously active *cytomegalovirus* (CMV) promoter and partial IL4 cDNA/IgE receptor α chain, followed by a SD and an intron sequence. 3' to this gene cassette an ATG-less human placental AP with its endogenous SA but lacking the secretion signal is included and serves as a reporter. In the wild-type situation, splicing will lead to an active AP, which is easily detectable in vitro. The introduction of a SA vector in front of the AP cassette will suppress AP activity, if a strong SA has been used. Thus, the strength of splice sites can be pretested in vitro. A recent survey of EST sequences in the public database has revealed alternative splicing to be more a rule than an exception. Concerning the gene-trap screen, this means that there might not be a final solution to the splicing-around problem. Focusing on transcriptional termination sequences to prematurely terminate trapped transcripts may turn out to be the way to go. Transcriptional termination consensus sequences have been described and tested in vitro. However, the 3' UTR is suspected to interact with the splicing machinery itself. As Barabino and Keller put it: "why does 3' end processing require a 1 MDa multicomponent machinery, if not to establish a network of weak cooperative interactions?" In case strong polyadenylation sites are capable of activating cryptic splice sites within the reporter–selector cassette, false negative expression results may be observed at least in specialized cell types.

Genotyping Mice with a Gene-Trap Mutation

Identification of heterozygous mutants

F1 generation mutants are easy to genotype by using probes or primers recognizing vector sequences. Dot blot analysis or a genomic PCR can be utilized to distinguish between positive and negative offspring and to give an estimate on copy number in the case of multiple integrations. In case the trapped gene is expressed in the tail tip, β-gal staining of tail biopsies have been applied successfully.

Identification of homozygotes using quantitative hybridization

Genotyping F2 gene-trap mice is a more laborious task. Internal probes or primers are not appropriate to distinguish between

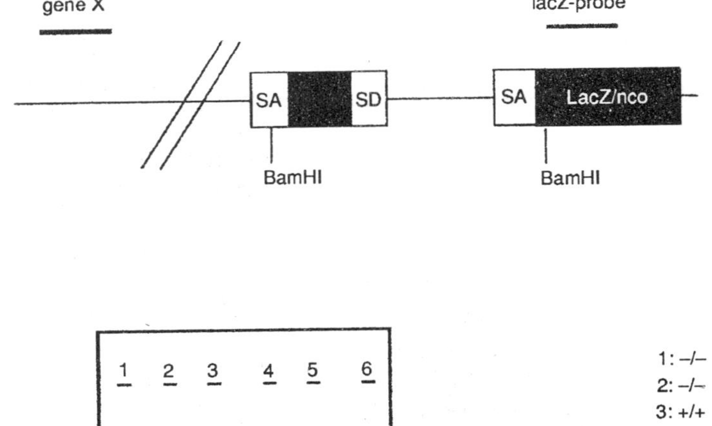

Fig. 5.3. In a quantitative experiment, a vector and a non-vector-related probe is used to estimate the copy number of the gene-trap vector integrations in the F2 generation.

heterozygotes and homozygotes, unless an unrelated genomic probe is used as a loading control.

Identification of mutants using restriction fragment-length polymorphisms

Often, cDNA, EST fragments, or PCR products are available, which can be used as external probes. In these cases, *restriction fragment length polymorphisms* (RFLPs) can generally be found by utilizing restriction enzymes that cut within the gene-trap vector.

Identification of mutated loci by plasmid rescue

Quantitative hybridization is subject to interpretation and, therefore, cannot be the method of choice. Some researchers take advantage of the bacterial ori and amp to use a plasmid rescue in order to isolate flanking probes, which are outside to the integrated gene-trap vector. It is recommended to check for a retainment of bacterial sequences prior to plasmid rescue by PCR or dot blot methods. Araki et al. have reported the loss of the ampicillin resistence cassette in more than 60% of examined cases. Probably due to homologous recombination between the LTRs when utilizing retroviral trap-vectors, only 20–30% of examined clones lose the bacterial sequences.

In order to perform plasmid rescue, enzymes need to be found, which will release genomic fragments of reasonable length. Genomic fragments over 20 kb have been cloned successfully. For transformation, it is recommended to use a bacterial strain with a defective DNA recombination system in order to be able to propagate repetitive sequences that are frequent in intronic DNA. A highly recommended strain is the *stbl2* bacterial strain, which is suitable for the cloning of unstable inserts such as retroviral sequences or direct repeats.

Identification of mutants by inverse PCR-generated probes

In our hands, the most successful method by far to generate external genomic probes for yet uncharacterized loci is inverse PCR.

Identification of trapped loci by adapter-mediated PCR

Recently, an adapter-mediated PCR approach to generate an external probe has been proposed. The protocol is straightforward in order to genotype mice with the scrambler mutation, but can be adapted on genotyping gene-trap mutations. Briefly, the protocol involves the digestion of genomic DNA, anchoring with a blunt-end adapter, and subsequent PCR with both adapter-specific and gene-specific primers.

Identification of homozygosity by breeding

In case mutant mice are viable and fertile, a reliable way to verify homozygosity is achieved by breeding potential homozygotes to wild-type mice. In case of homozygosity, naturally each of the offspring will be heterozygous for the trap mutation.

Accomplishments of the Gene-Trap Technology in the Mouse

Large-scale insertional mutagensis is performed by a number of laboratories worldwide and has already led to an attractive collection of trapped loci, which are being made available to the scientific community on both a profit and a nonprofit basis. Within the German gene-trap consortium, we have trapped 14,000 ES clones, of which by now more than 7000 have been identified by 5' RACE already. More than 1000 of the trapped loci are yet unknown, and more than 900 carry mutations in previously identified mouse ESTs. Overall, more than 2000 hits show homology to independent mouse or human unigene clusters, and more than 300 traps have links to the *Online Mendelian Inheritance in Man* (OMIM) database and are, therefore, human disease-related. Even mutagenesis with alkylating agents is never completely random. Chowdhury et al. have demonstrated more or less stochastic ntegration for 55 integrations of both a IRESβgeo and a SAβgeo vector into the mouse genome. To show that there is no preferred locus of integration, we plotted gene-trap hits in our database on the corresponding human chromosomes. The vectors that have been used are the PT1βgeo and the retroviral U3βgeo. Three hundred-eighty-two hits have links to the radiation hybrid mapping panel and can thereby be allocated to the corresponding genomic region. This analysis has revealed a more or less random integration of the different gene-trap vectors into the genome, taking into account the individual size of the chromosome.

Nevertheless, "*hot spots*," with more than 15 integrations having an e-value less than 10^{-10}, are being observed. Examples are "*Bruce*" (17 hits) and the "*Jumonji*" locus (30 hits). Interestingly, *Bruce* has been trapped by Bill Skarnes' group and *Jumonji* has also been trapped independently by two other groups. Another example is *"R-PTP-κ,"* which has been trapped once by Salminen et al., Skarnes et al., and twice by our gene-trap consortium. Each group has used different, but more or less related, trap-vectors. Whether this phenomenon is based on the individual expression level, chromatin status, or vector homology is under investigation.

Since the publication of the first 60% of the human genome sequence, previously "*unknown*" trapped loci show homology to the unfinished human genome sequence tags, emphasing the use of gene-trapping as an appropriate tool for gene discovery.

In summary, the gene-trap technology fullfills minimum requirements for a large-scale functional analysis of the mouse genome:

1. Gene-trap vectors randomly integrate into the genome.
2. The site of vector integration can be determined.
3. The reporter gene mimics endogenous gene expression.
4. The tagged gene is most likely mutated by gene-trap vector integration.
5. The mutation can be transmitted through the germ line using the ES cell technology.

Since both the production of ES clones and the identification using RACE PCR methods have been already semiautomated within the German gene-trap consortium, we will be able to saturate the mouse genome with gene-trap vector integrations within the next 5 yr.

Some Future Aspects

Modifications of trapped loci

From many different areas, there is constant influence on the gene-trap technology. On the one hand, a new generation of vectors is on the rise, which will allow also somatic mutations of gene-trap integrations at any given time point or in a given tissue by using consensus or modi-fied Cre/loxP or Flp/frt systems. The use of modified recombination sites not only allows to rescue gene-trap phenotypes and, thereby, demonstrate a causal relationship of phenotype and genotype, but also to irreversibly bring new genes under the control of the promoter of a trapped gene. Gene-trap vectors with modified loxP sequences are already in use. The use of loxP or FRT sites in gene-trap vectors will eventually saturate the mouse genome with these sites, which will allow the introduction of large chromosomal rearrangements after breeding two independent gene-trap lines.

Nuclear Transfer

We see a second major improvement of gene-trapping within the nuclear transfer of ES cells and somatic cells. A major limitation of the gene-trap technology is the generation of chimeras and breeding for germ line transmission. Applying nuclear transfer technology to the trapped ES clones will speed up germ line transmission.

Recently, it has been shown that transgenic mice can be generated by replacing the nucleus of an unfertilized oocyte with the nuclei of wild-type and even targeted ES cells. Embryonic development is subsequently triggered by treatment in strontium chloride. Interestingly, it turns out that cells in G0/G1 are favored for nuclear transfer techniques.

Since 35–40% of ES cells, when maintained under standard conditions, are in the S-phase, a prerequisite of cloning ES cells is a cell cycle arrest. Low serum conditions are, therefore, facilitating the cloning of ES cells. Although the authors succeeded in cloning of cells carrying a targeted mutation, their data also prove the difficulties of cloning higher passages of ES cells. The reason for this fact might be the loss of epigenetic markers associated with imprinted genes during culture, which has been described by Dean et al. for high-passage ES cells.

In addition, the technique does not seem to be applicable on pure genetic 129 background cells: the authors report F1 outbred cells of 129 × C57 to be most efficient. Therefore, the nuclear transfer is unlikely to be applicable on existing gene-trap clones, which are mostly 129-derived. This aspect needs to be considered in the future. By using the cloning technique, a successful generation of transgenic mice will not be dependent anymore on the maintenance of an undifferentiated status of ES cells, since it has been demonstrated that differentiated cells are suitable for cloning as long as a normal karyotype is maintained. The generation of mice with targeted or gene-trap insertions by nuclear cloning and, thereby, circumventing the chimera stage is feasable and will save a considerable amount of both time and money in the future. It will make the gene-trap technology even more suitable for high-throughput mutagenesis screens.

DNA chip technology

The third major improvement currently arising is the DNA chip technology. As a systematic approach to identify expression profiles of novel genes and to characterize biologically active substances, microarrays have already proven to be extremely valuable in complementing the genome sequencing effort. The majority of genes are still unknown, and whole genome expression profiling will provide information on the expression levels and patterns of large numbers of ESTs and unknown genes at the same time. DNA microarrays, containing sequence information of trapped genes, will combine the expression profiling with the gene-trap technology, to be able to identify differentially expressed genes for which mutant ES cells exist and mouse models can be established. Joint efforts are already being undertaken to achieve this exciting goal.

A worldwide complementation of the sequencing effort is already underway in order to fill in the "*phenotype gap*". Large numbers of gene-trap mutants are currently under analysis and will contribute considerably to functional genomics in the near future. What is required is a worldwide coordination of the different activities to a joint effort to avoid redundancy in the future.

As a first step towards this goal, gene identification data by gene-trapping can be submitted to the NCBI Genbank. A number of laboratories applying the gene-trap technology use their own nomenclature, which is confusing to the scientific community. A common nomenclature for gene-trap mutations is urgently needed. Expression data of genes are made available through the mouse expression database, which can be linked to gene-trap libraries. Selected databases of gene-trap screens are already accessible in the World Wide Web.

In order to perform a large-scale phenotypical analysis of gene-trap mutants without overt phenotypes, phenotype centers should be established, which coordinate standardized quantitative analytical methods as well as noninvasive tests of function. This includes the examination of clinically relevant parameters such as clinical–chemical and biochemical parameters, dysmorphology including pathological assessment, neurological analysis, and behavioral determination. A demonstration of handling large numbers of mice with a standardized phenotyping protocols is currently given by the ENU mutagenesis project and needs to be applied on the gene-trap screen as well.

Materials

Recovery of Early Postimplantation Embryos

1. 1 Pair of coarse forceps.
2. 1 Pair of fine scissors.
3. 2 Pairs of fine forceps.
4. 2 Pairs of watchmaker's forceps.
5. Pasteur pipets.
6. Pasteur pipets with "wide opening."
7. Phosphate-buffered saline (PBS).
8. Petri dishes.
9. Dissecting microscope.

β-Gal Staining of Cultured Cells, Whole Embryos, and Tissues

1. 10X PBS: 80 g NaCl, 2 g KCl, 14.4 g Na_2HPO_4, 2.4 g KH_2PO_4 to 800 mL with H_2O, HCl to pH 7.6, H_2O to 1 L.
2. Solution A: Kanolinite phenylphosphonate (KPP): 100 m*M* potassium phosphate buffer, pH 7.4, store at room temperature.
3. Solution B: 0.2% glutaraldehyde (GDA) in solution A containing 5 m*M* EGTA and 2 m*M* $MgCl_2$ (store at –20°C).
4. Solution C: 0.01% Na-desoxycholate and 0.02% Nonidet P-40 (NP40) in solution A containing 5 m*M* EGTA, and 2 m*M* $MgCl_2$ (store at room temperature).
5. Solution D: 0.5 mg/mL 5-bromo-4-chloro-3-indolyl-β-D-galactopyranoside (X-gal), 10 m*M* $K_3[Fe(CN)_6]$, and 10 m*M* $K_4[Fe(CN)_6]$ in solution C (store at –20°C in the dark).

In Vivo Staining of ES Cell Colonies

1. Fluorodeoxyglucose (FDG) or Imagene.
2. Loading medium for FDG: dilute FDG stock solution (20 m*M* FDG in 10% *dimethyl sulfoxide* [DMSO]) 1:10 with sterile water; a 1:1 mixture of tissue culture medium and the 1:10 diluted FDG solution is used as loading medium. FDG stocks may vary depending on batch and supplier.
3. Fluorescence microscope.
4. ES cells grown on tissue culture dishes.

Paraffin Embedding of β-Gal-Stained Tissues

Material

1. Paraffin for histology.
2. Isopropanol.
3. Screw cap tubes.
4. Casting mould.
5. Small spatula.
6. Microtome and holder for fixing the paraffin block to the microtome.
7. Xylene.

Counter Staining Sections of lacZ-Stained Material

1. Histological staining trays.
2. Hematoxilin and eosin solution.
3. 60, 80, 96, and 100% ethanol.
4. Xylene.
5. Embedding medium.

ES-Cell Culture Media and Solutions

1. PBS without Mg^{2+} and Ca^{2+}.
2. Gelatinized plates.
3. Geneticin (G418) stock 200 mg/mL (4°C).
4. 0.05% Trypsin solution in Tris-saline-EDTA buffer (ICN Flow).
5. Gelatin solution: 0.1% in *aq* (Fresenius)
6. 2X concentrated freezing medium ES-Zellen: 50% fetal calf serum (FCS), 20% DMSO.
7. Nucleosides 100 ml 100X stock solution: 80 mg Adenosine, 73 mg Cytidine, 85 mg Guanosine, 24 mg Thymidine, 73 mg Uridine, dissolve in 100 mL H_2O (pre-warm to 37°C and filter-sterilize. Aliquots (e.g. 5 ml in 15 ml Bluecap) can be stored at 4°C. Precipitate re-dissolves at 37°C.

tbv-2 ES medium

Dulbeccos' Modified Eagles Medium (DMEM), high glucose, without Na-pyruvate containing: 15% FCS (heat-inactivated for 30 min at 56°C), 1 m*M* Na-pyruvate, 2 m*M* glutamine, $10^{-4}M$ β-mercapto-ethanol, use up within 2 wk. Add before use: 1000 U/mL *leukemia inhibiting factor* (LIF).

R1 ES-medium

DMEM, high glucose, without Na-pyruvate containing: 20% FCS, 1 m*M* Na-pyruvate, 2 m*M* glutamine, $10^{-4}M$ β-mercaptoethanol, use up within 2 wk. Add before use: 1000 U/mL LIF, 5.7 mL nonessential amino acids; optional: 50 U/mL Pen–Strep.

CJ7 ES-medium

DMEM, high glucose, without Na-pyruvate containing 15% FCS, 2 m*M* glutamine, 5.7 mL nonessential amino acids, $10^{-4}M$ β-mercaptoethanol, 5 mL nucleoside (100X), 1000 U/mL LIF; optional: 50 U/mL Pen–Strep.

Electroporation of ES Cells ("Gene-Trap Conditions")

1. ES cell medium.
2. PBS without Mg^{2+} and Ca^{2+}.
3. 10-cm gelatinized plates.
4. G418 stock (1000X) 200 mg/mL (4°C)
5. Electroporation apparatus.
6. Electroporation cuvets.

Infection of ES Cells with Retroviral Vectors

1. Gelatinized tissue culture plates.
2. Medium-containing retrovirus.
3. Polybrene.
4. ES cell medium containing LIF- or *Buffalo rat liver* (BRL)-conditioned medium.

Cloning Flanking Sequences by Inverse PCR

1. Genomic DNA carrying the transgene insertion.
2. Oligonucleotide primers.
3. Restriction enzymes.
4. T4 Ligase.
5. *Taq* DNA polymerase, appropriate reagents and equipment for PCR.

Novel PCR-Based Technique of Genotyping Applied to Identification of Scrambler Mutation in Mice

1. Microcentrifuge.
2. Minigel electrophoresis system.
3. UV transilluminator.
4. Thermal cycler.
5. Agarose.
6. Proteinase K (DNase-free).
7. *Rsa*I and *Hae*III restriction endonucleases in supplement with the buffers.

8. T4 DNA ligase in supplement with the buffer. BioTherm DNA polymerase in supplement with the buffer DNA polymerization mixture (20 m*M* dNTPs).
9. 100-bp DNA ladder.

Methods

Recovery of Early Postimplantation Embryos

1. Sacrifice the pregnant female by cervical dislocation.
2. Lay mouse on its back, make a large V-shaped incision into the skin with the tip of the V just anterior to the vagina.
3. Fold the skin back, cut the abdominal wall similarly to the skin and fold back.
4. Find the uterus, hold the uterine horns with coarse forceps at the cervical end and cut at the uterine-cervical junction.
5. Pull uterine horns slightly, trim away the mesometrium (part of the broad ligament that is attached to one side of each uterine horn) along the uterine wall, and cut off the uterine horns at their anterior ends.
6. Put uterine horns into Petri dish containing PBS.
7. Cut the uterine horns between the decidual swellings.
8. Under the dissecting microscope, tear the uterine muscle on the opposite side to the attached mesometrium with fine watchmaker forceps.
9. Free the decidual swelling from the uterine wall by holding the muscle with one fine watchmaker forceps and sliding along between the torn muscle and deciduum with the second fine watchmaker forceps.
10. Transfer deciduas into fresh dish containing PBS.
11. Hold the deciduum on the mesometrial (broad) end with one fine watchmaker forceps. Insert the point of the closed second forceps in the midline above the redish streak (which is the embryo), through the deciduum, and open fine watchmaker forceps splitting the deciduum.
12. Grasp the split parts and pull apart. The embryo, surrounded in membranes, usually remains attached to one decidual half.
13. Gently push the embryo with the tip of the closed watchmaker forceps until the embryo and its membranes are entirely free.
14. Transfer the embryo with a Pasteur pipet (6.5 and 7.5 days postcoitum [dpc]) or a wide opening Pasteur pipet (8.5 and 9.5

dpc). Older specimens can be transferred by "*scooping*" them with a curved forceps.

Removal of Reichert's membrane (6.5–8.5 dpc)

Grasp Reichert's membrane at the extraembryonic portion of the egg cylinder (away from the embryo) with both watchmaker forceps and gently tear it open. Most of the membrane can be torn off leaving behind only some remnants at the ectoplacental cone that do not impair staining and analysis of the embryo. It is advisable to leave the membranes attached to the embryo to discover extra-embryonic staining as well.

Removal of extra-embryonic membranes (>d 8.5)

Grasp visceral yolk sac with both forceps. Tear until the embryo is freed from the yolk sac but still connected with it by the umbilical cord. Hold the cord with one forceps and tear off the yolk sac distally with second forceps. If the amnion, a very thin cellular membrane, is still surrounding the embryo remove that analogously to the yolk sac.

β-Gal Staining of Cultured Cells, Whole Embryos, and Tissues

1. Wash in PBS: for cells: aspirate the medium and replace with PBS; repeat. For embryos: transfer into PBS, gently swirl around. For tissues: transfer into PBS, gently swirl around.
2. Fix in buffer B: for cells: add sufficient buffer B to the plate such that the cells are well covered and leave for 5 min at room temperature. For embryos: up to d 9.5, add 1 mL buffer B for 10–20 embryos and leave for 5 min at room temperature. For d 10.5 to 12.5 embryos, add 5–10 mL buffer B for 10 embryos and leave for 15 min at room temperature. For tissues: add about ten times the volume of the tissue of buffer B and leave 15–60 min (depending on size) at room temperature.
3. For all fixation steps: aspirate well wash buffer before adding buffer B to prevent dilution.
4. Wash 3 times with 10 mL buffer C at room temperature: for cells: 5 min each. For embryos up to day 9.5, 5 minutes each. For embryos up to d 10.5–12.5: 15 min each. For tissues: 15–60 min (depending on size).
5. Replace buffer C with buffer D and incubate at 37°C. Before adding buffer D, aspirate well buffer C: for cells: add sufficient buffer to the plate such that cells are well covered and that the solution will not evaporate: for embryos up to d 9.5, add 1 mL

buffer; for 10–20 embryos, for d 10.5–12.5, add 5–10 mL for 10 embryos. For tissues: add about ten times the volume of the tissue.

6. After staining, wash samples 3 times in 10 mL buffer C.
7. Samples can be stored for short term (a few days) in solution C at 4°C, but for prolonged storage, the specimens should be fixed again in 4% paraformaldehyde for 2 h at room temperature and kept in 70% ethanol at 4°C.

In Vivo Staining of ES Cell Colonies

1. FDG: aspirate tissue culture medium and add sufficient loading medium to cover the cells (i.e., 1 mL/30-mm dish, 2 mL/60-mm dish, 3 mL/90-mm dish). Incubate for 1 min. Change back to regular tissue culture medium. Imagene: add dye directly to the medium at a final concentration of 33 μM.
2. Incubate at 37°C for 1 h (FDG) or 2 h.
3. Identify fluorescing colonies with a fluorescence microscope using 10× or 20× objectives and filters for fluorescein. Mark the positions of positive clones with a dot on the bottom of the dish.

Paraffin Embedding of β-Gal-Stained Tissues

1. Place the dehydrated sample in 10 mL 100% isopropanol for 2 h with one change of the isopropanol at room temperature. Alternatively, samples can be dehydrated directly in isopropanol similarly as given below for ethanol. Using isopropanol, the steps are 50, 75, 90%, and 2 times 100%.
2. Preinfiltrate with paraffinisopropanol (1:1) at 60°C.
3. Infiltrate with paraffin at 60°C.
4. Place into prewarmed (60°C) mould and orientate the specimen with a needle or spatula and fill mould with paraffin.
5. Depending on the mould used, directly cast the holder for fixing the paraffin block to the microtome. After hardening, remove the paraffin block from the mould, and prepare 10 μM sections.
6. Dewax sections 1 to 2 min in xylene. Sections can now be embedded or processed for counterstaining.

Counter Staining Sections of lacZ Stained Material

1. Submerge sections for 1 min in hematoxilin solution in staining tray.
2. Rinse through staining tray for 5 min with tap water.
3. Submerge sections for 2 min in eosin solution.
4. Rinse 5 min with distilled water.

5. Dehydrate frozen sections and paraffin sections in ascending ethanol (60, 80, 96%, and 2X 100%, 2 min each).
6. Remove the ethanol by two incubations in xylene (1 min each), add Eukitt, and put coverslip on. Any other commercially available embedding solution can be used. Methacrylate sections can be dried and embedded directly.

Electroporation of ES Cells ("Gene-Trap Conditions")

Expansion of ES Cells

1. Thaw one vial of early-passage ES cells, wash as usual, and plate the cells on a 60-mm gelatin-coated Petri dish with primary embryonic feeder cells (EMFI). EMFI cells should be confluent on the plate.
2. Change the ES cell medium once each day.
3. After 2 d, trypsinize and expand the cells on one to two 90-mm gelatin-coated Petri dishes (with EMFI cells) dilution 1:3 up to 1:8, depending on the cell density.
4. Wait another 2 d and transfer the cells to fresh feeder plates again dilution 1:3 or 1:4.
5. After 2 d expand the cells to at least 2 × 200 mm or 6 × 90 mm gelatinized Petri dishes (only a dilution of 1:2) and start the transfection 36 h later in one electroporation cuvette. The cell number on 2 × 200 mm subconfluent Petri dishes should be approx 1 × 10^8 cells. Change medium in the plates 6 h before the electroporation.

Electroporation and Selection

1. Trypsinize ES cells for 10 min in 3 mL trypsin/dish, and pipet them gently up and down to aqcuire a single-cell suspension, add 3 mL medium/dish, and transfer the cells to 2 Falcon tubes.
2. Centrifuge the cells for 3 min at 270*g*.
3. Resuspend the cells in 10 mL PBS.
4. Dilute an aliquot 1:10 and count the cells (keep the cells on ice).
5. Centrifuge 1 × 10^8 cells 5 min at 270*g* for one cuvette.
6. Add to the pellet 500 μL cold PBS and 100 μL of the DNA (120 μg), no more than 700–800 μL in total (Vector linearization: digest with adaequate enzyme 1 U/μg for at least 4 h, phenol-extract, and precipitate the DNA at 70°C for 10–15 min. Wash in 70% ETOH and air-dry under sterile conditions).

7. Transfer the suspension to the electroporation cuvette.
8. Set up the electroporation conditions in advance (0.8 kV, 3 μF for the Bio-Rad gene pulser).
9. Transfer the cuvette into the cuvette-holder with electrodes facing the output leads and deliver electric pulse.
10. Remove the cuvette from the cuvette-holder and leave it at room temperature or on ice for 10–20 min.
11. Transfer the cell suspension from one cuvette into 12 mL ES cell medium. Seed the electroporated cells at a density of 2.5–5 $\times$ 10^6 cells/90-mm dish on a gelatinized plate in medium containing LIF. The cell concentration per plate must be adjusted depending on the vector such that no more than 200–500 neo^R colonies are obtained on each plate.
12. Change the ES medium the next day.
13. Two days after electroporation, add the drugs for selection to the ES medium (e.g., G418:200 μg/mL [active]; puromycin: 1 μg/mL).
14. Change the selection medium every day for the first 3 d, than every other 2 d.
15. About 6–8 d of selection, drug-resistant colonies should have appeared.
16. After 8 to 9 d of selection, colonies are picked and plated on 96-well feeder plates containing ES medium. Stop the selection-pressure and exchange the ES cell medium each day. After 1 d of growth, tryplate the colonies, after another 2 d, trypsinize and dilute the clones 1:3. After another 2 d, split each 96-well replica-plate 1:1 on 2 $\times$ 48 well feeder plates.

Infection of ES Cells with Retroviral Vectors

1. Plate ES cells on gelatinized tissue culture dishes at a density of 3 $\times$ 10^6 cells/90-mm dish in ES cell medium supplemented with LIF- or BRL-conditioned medium.
2. After 24 h, aspirate medium, add 5 mL fresh medium containing the retroviral particles at a multiplicity of infection (moi) <1 and 5 μg/mL polybrene to obtain single integration per clone.
3. After overnight culture (14 h) remove virus-containing medium, add 10 mL fresh medium, and culture for an additional 24 h.
4. Change the medium to selection medium and change medium every other day. Drug-resistant colonies should become visible between 7–10 d.

Generation of ES Cell-Chimeric Embryos for lacZ Expression Analysis

1. Inject 10–15 albino-derived blastocysts with ES cells derived from a pigmented mouse for each early embryonic stage to be analyzed.
2. Inject an additional 10–15 blastocysts for the d 12.5 control.
3. Transfer the embryos to foster females in groups of 10–15 (1 foster/stage to be analyzed, and 1 additional for the control).
4. Recover embryos at desired stages and stain for β-gal activity.
5. Recover control embryos at d 12.5, and monitor the embryos for pigmented cells in the eye (which can be easily seen by the naked eye or under a dissecting microscope), and stain for β-gal activity.

Cloning Flanking Sequences by Inverse PCR

1. Digest genomic DNA to completion with a restriction enzyme that does not cut within vector DNA sequences between oligonucleotides 1 and 2, and that cuts at least 0.5–1 kb outside the vector in the genome flanking DNA. Heat-inactivate (15 min at 65°C) the restriction enzyme when digestion is complete.
2. Ligate the digested DNA with 8 Weiss U of T4 DNA ligase at 14°C for 16 h at a concentration of 1 μg DNA/mL in a total volume of 600 mL. This DNA concentration favors intramolecular as opposed to inter-molecular ligation. Heat-inactivate (15 min at 65°C) the ligation mixture.
3. Linearize DNA with appropriate restriction enzyme. Add the enzyme and its 10X incubation buffer directly to the ligation reaction. After appropriate incubation, heat-inactivate (15 min at 65°C) the restriction enzyme.
4. Precipitate the DNA in high salt and isopropanol in the presence of 10 μg/mL tRNA or 1 μg/mL glycogen.
5. Set up the amplification reaction by combining:
   (a) 1–100 ng genomic DNA (cut, ligated, and recut).
   (b) 5 m*M* dNTPs each.
   (c) 1 to 2 U *Taq* DNA polymerase.
   (d) 0.1–0.5 μ*M* primers each.
   (e) 5 μL 10X *Taq* incubation buffer.
   (f) H_2O to a total vol of 50 μL.
6. 20 microliters of each test reaction is analyzed on a 1% agarose gel. A band corresponding in size to the fragment deduced from primer placement and genomic Southern blot data should be visible

on an ethidium bromide-stained gels. Confirm that the correct piece of DNA has been amplified by Southern blot analysis of the gel by hybridization with vector sequences contained in the amplified fragment.

7. Digest the rest of the test reaction, or additional reactions, with restriction enzymes, which lie outside the first primer pair but within vector sequences. Alternatively, PCR primers that contain internal restriction sites can be designed. Analyze an aliquot of this digestion on an agarose gel. The amplified fragment should have shifted downward by the anticipated number of base pairs.
8. Purify the DNA fragment from an agarose gel (by electroelution, low melting point agarose, etc.).
9. Ligate the digested and purified PCR fragment to a plasmid vector cut with the same enzymes. If no convenient restriction enzymes (which do not cut within the genomic flank) exist, clone the purified PCR product blunt-ended into a plasmid vector, or use appropriate plasmids for the direct cloning of PCR products.
10. Transform competent bacteria with an aliquot of the ligation reaction and screen transformants for the presence of the desired clone.
11. The efficiency of the first steps of the protocol can be estimated on agarose gels. If multiple bands are visible after the PCR step, try less input DNA or raise the annealing temperature above the melting temperature (T_m).

Novel PCR-Based Technique of Genotyping Applied to Identification of Scrambler Mutation in Mice

Genomic DNA extraction

1. Place a cut tip of mouse tail into 0.5 mL of lysis buffer (100 m*M* Tris-HCl, pH 8.5, 5 m*M* EDTA, 0.2% SDS, 200 m*M* NaCl, 200 mg/mL proteinase K).
2. Shake overnight at 55°C.
3. Centrifuge for 20 min at 15,000*g*, at 4°C.
4. Collect supernatant and add 0.5 mL of isopropanol. Mix gently. Genomic DNA should become visible as a fibrous substance.
5. Centrifuge for 2 min at 15,000*g*, and discard the supernatant.
6. Add 1 mL of 80% ethanol, centrifuge for another 2 min, and discard the supernatant.
7. Repeat this washing with 80% ethanol.

8. Dry the pellet and dissolve it in 100 mL of Tris-HCl, pH 8.0. Normally, this requires 2 h of shaking at 55°C.

Endonuclease digestion

1. Take 5 μL of genomic DNA solution as described above. Set the digestion in the total vol of 50 μL using 15 U of an enzyme that will generate blunt ends.
2. Incubate for 2 h at 37°C.
3. Extract digested DNA with phenol:chloroform and precipitate with sodium-acetate and ethanol.
4. Dissolve in 10 μL of 10 m*M* Tris-HCl, pH 8.0.
5. Run 1 μL of this solution on 1.5% agarose gel to estimate the quantity: each lane should contain 50–100 ng of DNA.

Adapter ligation

1. The long strand of pseudo-double-stranded adapter was: 5'-AGCAGCGAACTCAGTACAACAACTCTCCGACC-TCTCACCGAGT-3'.
2. The short strand was: 5'-ACTCG- GTGA-3'.
3. Perform the adapter ligation reaction in the final vol of 10 μL. The mixture should contain 200–400 ng of DNA, 2 μM of each adapter strand, and 1 U of T4 DNA ligase in the reaction buffer provided by the enzyme supplier.
4. The reaction is carried out for 3 h at room temperature or overnight at 12–16°C.

PCR conditions

1. Primers used for PCR amplification of the scrambler-mutation were:

   (a) Sc1: 5'-TTTTGTCCTTCTCTATAACT-3'.

   (b) Sc2: 5'-CCTGGGA-TAATGGGGTAAG-3'.

   Instead, nested primers can be designed according to the gene-trap vector sequence. Annealing temperatures and extension times have to be determined individually.
2. The distal adapter primer (DAP): 5'-AGCAGCGAACTCAGTACA ACA-3' (corresponds to the 5' part of the long strand of adapter).
3. Add to the ligation mixture: 4.5 μL of 10X PCR buffer, 10 nmol of each dNTP, 15 pmol of Sc1 primer, and water to the final vol of 45 μL, and cover the mixture with mineral oil.
4. To perform the "*hot start*," heat to 94°C in the PCR machine. Add 2.5 U of BioTherm DNA polymerase in 5 μL of 1X PCR

buffer to the reaction mixture, while avoiding a cooling of the tube.

5. Carry out 22 cycles of PCR: 10 s at 94°C, 30 s at 55°C, and 60 s at 72°C.
6. When the amplification is complete, dilute the PCR mixture 40-fold with water.
7. Transfer 1 μL of this dilution into 25 μL of the second PCR mixture. The latter contains PCR buffer components, 5 nmol of dNTPs, 8 pmol of Sc2 primer, and 5 pmol of DAP, 2 U of BioTherm DNA polymerase (which should be added directly to the cold mixture). Run 23 cycles of PCR: 10 s at 94°C, 30 s at 60°C, and 60 s at 72°C (please note that the annealing temperature and extension time is given for the scrambler mutation).

Analysis of PCR-Amplified Fragments

Separate PCR products by horizontal electrophoresis through a 1.5% agarose gel in parallel with the DNA marker for size estimation. To visualize the DNA, stain the gel with ethidium bromide.

In Vivo Staining of ES Cell Colonies

Verify positive clones afterwards by X-gal staining. The fluorescence signal of a true positive clone can vary tremendously, and most of the signals localize to just a portion of the colony. Use a cell line known to express *lacZ* at clearly detectable levels as a positive control.

Notes

β-Gal staining of cultured cells, whole embryos, and tissues

1. Always prepare fresh fixing solution before use or store at –20°C; other fixations can also be used (e.g., 2% glutaraldehyde or 4% paraformaldehyde [PFA] are possible).
2. When solution D is prepared freshly chill on ice for 10 min, spin down precipitate, and aliquot supernatant.
3. The staining solution D can be reused several times; filter after each use, and keep in dark at –20°C.
4. A 50 mg/mL X-gal stock solution (=100×) can be made both in DMSO or dimethylformamide (DMF). Keep at –20°C. DMF has the advantage to stay liquid at –20°C.
5. $K_3[Fe(CN)_6]$ and $K_4[Fe(CN)_6]$ should be kept as 0.5 *M* stock solutions (=50X) in dark bottles at (–20°C).

6

BIOREACTORS

Historically, cell cultivation was first devised at the beginning of the 20th century. For many years, it was restricted to simple small-scale propagation systems applied to tissue culture for basic research. Simultaneously, large bioreactors were already in use for the production of commercially interesting secondary metabolites exploiting microbial fermentation. The development of industrial-scale cell culture bioreactors started in the mid-1950s in response to the need for mass cultivation techniques that were suitable for vaccine production as demanded by the massive vaccination programs launched at that time. Previously, roller bottles were used to culture primary cells at relatively limited scale for vaccine production. These initial cell culture bioreactors were speci.cally designed for adherent cells, examples are plate propagators and packed beds. The first commercially interesting suspension cell products [food-and-mouth disease vaccine produced in suspension culture of BHK cells and interferon produced in Namalva cells] stimulated the adaptation of homogeneous bioreactor systems used for microbial culture to the requirements of the mechanically more sensitive animal cells. The advent of the monoclonal antibody era in the 1970s then gave rise to the development of a plethora of different bioreactors and culture systems suitable for suspension cell culture with special emphasis on increasing the product yield per unit volume through improved nutrient supply and waste product removal. These specialized systems include hollow fiber, fluidized bed reactors, and other different types of compartmentalized bioreactors based on cell immobilization and perfusion of fresh medium through the cell-containing compartment. The basic idea was to overcome the major limitations of cell cultivation, i.e., slow cell growth and low final cell densities, by providing an environment that allows the cells to continuously produce the product of interest at high levels. In parallel, a large number of cell retention devices for stirred tanks or airlift bioreactors were developed allowing for continuous exchange of medium in homogeneous systems. Recombinant DNA technology, which is the basis of modern biotechnology today, enabled the production of a series of protein therapeutics in mammalian cells in the 1980s. This novel opportunity had further impact on the development and optimization of bioreactors for suspension as well as anchorage-dependent cells in large scale. In this regard, a severe biomanufacturing bottleneck is forecasted for the coming decade due to the strong product pipeline for fully human monoclonal antibodies and ultra-large scale bioreactors beyond the 20 m^2 scale are discussed to meet this anticipated demand of several kilograms of protein per year. At the same time scaled-down systems are becoming more important to enable multiple parallel experimental approaches to be taken for cell line and process development. Since the mid-1990s, the available knowledge has also been applied to the design of bioreactors used for artificial organs and systems for tissue and stem cell culture. Other more recent applications are the production of viral vectors for genetic vaccination or gene therapy.

This chapter aims to review the different cell culture bioreactor systems developed for suspension and anchorage-dependent cells during the last 50 years of industrial cell culture. Major emphasis is given to well-established bioreactors of broad applicability and an attempt is made to give a comprehensive overview of cell culture systems developed for specific applications that in many cases might not be commercially available. The primary role of a bioreactor is to provide containment with suitable conditions for cell growth and product formation. In principle, this can be achieved by mimicking the environment a cell is exposed to in a tissue, i.e., ensuring hemostasis by exclusion of microorganisms, thermostatization, continuous supply of oxygen, nutrients, and growth factors, and removal of CO_2 and waste products. Adherent cells will also require a surface for attachment. Therefore, one way to describe bioreactors for animal cells is to categorize them into either suitable for suspension or anchorage-dependent cell culture. Some systems may also be modified or adapted for the cultivation of both types. In general, bioreactors can be distinguished in homogeneous systems where the cells are uniformly distributed in the bioreactor and heterogeneous systems, which are characterized by a separation of cells and medium using different types of membranes. Homogeneous systems are well-mixed systems whereas in heterogeneous systems gradient formation can occur.

The description of the different bioreactors covers functional principles, engineering considerations, and scale-up. Examples of applications in terms of cell lines cultured, products recovered, or novel purposes other than production of proteins are given. Advantages and limitations of the different systems are discussed as well as general performance regarding cell densities and productivity. Typical bioreactor operation modes are described in general terms and the relative merits for different applications are discussed. Finally, an attempt is made to summarize technical, scientific, economic, and regulatory considerations related to the selection of a certain bioreactor system and their impact on the design.

Bioreactors for Suspension Cell Cultures

Small-Scale Culture Systems

Small-scale culture systems are characterized by a relatively simple design and low level of instrumentation and control. Traditionally, roller bottles and spinner flasks have been used for small-scale suspension culture although even T-flasks, Petri dishes, multiwell plates, and other stationary culture systems are applicable for suspension cell propagation in small-scale. Spinner flasks are available from 125 mL to 5 L working volume for operation in humidified CO_2 incubators. Larger flasks up to 36 L require access to warm rooms or can be operated with heating belts. They are generally made of glass but flasks made of plastics are also available. The most primitive design consists of a flask with two side ports for inoculum and medium addition and a magnetically coupled stirrer either of a bar type on a central axis or a conical pendulum. The stirrer rate is maintained between 30 and 100 rpm, frequently as low as possible to prevent sedimentation of cells. The flasks are commonly kept in a humidified CO_2 incubator and gas transfer takes place via the headspace through slightly opened caps. Oxygen transfer coefficients in the rage of 0.1–4 h^{-1} are reported for spinners. A general drawback of this design is the very low oxygen transfer resulting in oxygen limitation already at relatively low cell densities. To maximize oxygen transfer in spinners the height to surface ratio should be kept low; i.e. the working volume should be minimized. To overcome this limitation but still keeping it a simple culture system a special floating stirrer has been suggested. Another solution to overcome gas transfer limitations is to submerge gas permeable membranes in the culture fluid. Heidemann et al. developed the so-called Superspinner where microporous polypropylene membranes are fixed on a pendulum stirrer in a standard 1 L Schott glass bottle and the incubator air/CO_2 mixture is pumped through the membrane and the headspace using a simple air pump. This spinner flask set-up is also used in a recently developed small-scale screening system where pH and oxygen electrodes are introduced for monitoring and control (DASGIP). Spinner flasks have been widely used for the cultivation of many different suspension cell lines (primary and transformed mammalian and insect cell lines) for a variety of applications within preclinical research, process development, and for inoculum expansion in production. In addition, shaker flasks are commonly used for the propagation of suspension cells, not only insect cells but also hybridoma or CHO cells. It is also possible to culture suspension cells in sterile plastic tubes; e.g., centrifuge tubes, placed in a modified shaker apparatus or roller drums inside a humidi- fied CO_2 incubator. This set-up is commonly used for simple screening purposes in media development trials where numerous parallel experiments are undertaken. Microtiter plates are increasingly used for screening and optimization purposes in cell culture, in the area of high-throughput protein expression, and biopharmaceutical cell and process development. However, little knowledge is currently

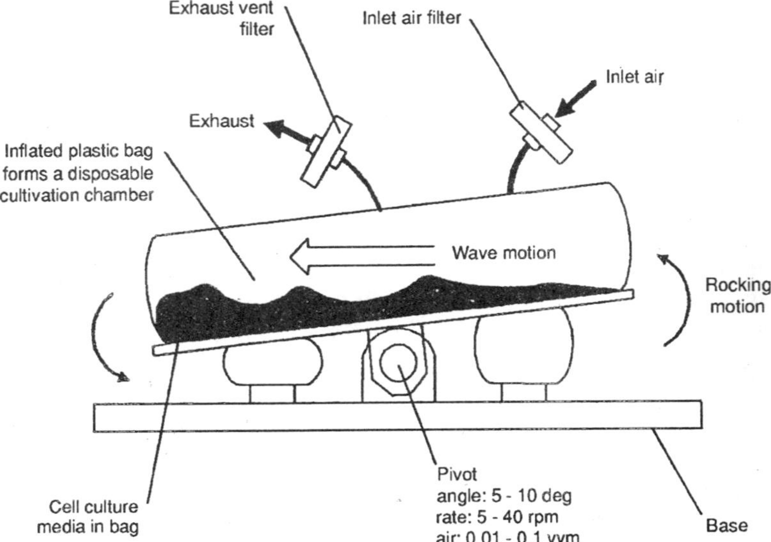

Fig. 6.1. Schematic representation of the wave bioreactor with wave-induced agitation.

available on the fundamental characteristics affecting culture conditions in this system. Generally, scale-down techniques are becoming more important in industrial process development and optimization. A review by Palomares and Ramirez summarizes the approach and provides examples and experimental configurations.

Another simple culture system is the WaveBioreactor made for suspension cell culture of 100 mL up to 500 L volumes in disposable plastic bags. Oxygen is transferred through the gas permeable wall. Due to the rocking motion of the bag on a rocker base oxygen transfer is considerably improved compared to spinner culture. The system can be operated in CO_2 incubators or standalone in combination with a heater and a CO_2 control unit. Singh describes the application of the system for monoclonal antibody production, adenovirus production with HEK 293 cells, and baculovirus production with insect cells. Due to the disposable nature of the bioreactor it offers clear advantages for cell therapy related to patient safety.

Stirred Tank Reactors

General

Stirred tank bioreactors (STR) are the most widely used bioreactor type to cultivate suspension cells, mainly due to the broad experience obtained in microbial fermentation. They have successfully been used for the cultivation of a wide variety of suspension cells and cells adapted to growth in suspension such as hybridoma cells, CHO, BHK 21, HEK293, and others. Commercial applications are the production of monoclonal antibodies, recombinant proteins such as blood coagulation factor VIII, tPA, erythropoietin, and other proteins for replacement therapy, vaccines, growth factors, and interferon. Novel applications are in gene and cell therapy such as the expansion of hematopoietic cells for reconstituting in vivo hematopoiesis in patients who have undergone intensive chemotherapy, the cultivation of human T-cells for immunotherapy, or for the production of virus vectors.

Major advantages are the relative ease of handling and the familiarity of technical personnel with this reactor type. Scale-up principles are better characterized compared to other bioreactor types and mechanical design principles for *sterilization-in-place* (SIP) and *cleaning-in-place* (CIP) are known to manufacturing engineers. Furthermore, regulatory agencies are experienced with products obtained from this bioreactor type in batch, fed-batch, and perfusion modes. Especially for contract manufacturers and pilot production facilities in the pharmaceutical industry the high flexibility in terms of applicable

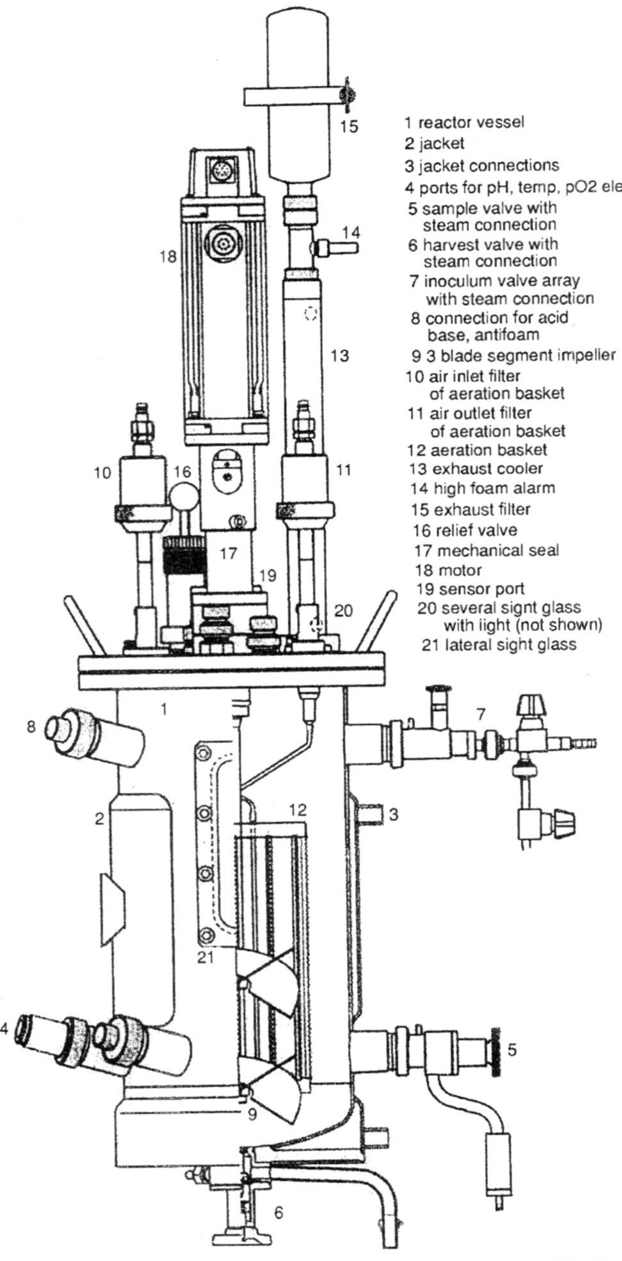

Fig. 6.2. Schematic drawing of stirred tank bioreactor equipped with silicone tubing for bubble free oxygen transfer.

working volumes, suitability for different cell types, operation modes, and products is a clear economic advantage.

Large-scale stirred tanks are used widely in the chemical and biochemical industries up to scales of several thousand cubic meters. For animal cells maximum working volumes of 15000 L have been reported. In the past some pharmaceutical companies have used their own engineering departments for the construction of large bioreactors. In addition, a number of manufacturers provide stirred tank reactors from 0.5 L to 15 m^3. During the last 15 years, a consolidation of the number of bioreactor manufacturers has occurred.

General design criteria are derived from STRs for microbial culture but modified to meet the requirements of the more sensitive animal cells. Specifically, the shear sensitivity of animal cells needs consideration regarding the design of impellers, use of baffles, aspect ratio, and oxygenation method.

Mixing

Agitation in stirred tanks aims for homogeneous suspension of the cells and the prevention of chemical (nutrients, waste products), physical (pH, oxygen, carbon dioxide), and thermal gradients in the vessel. Typically, large impellers (impeller diameter $\geq 0.5 \times$ vessel diameter) with an axial fluid flow characteristics such as marine-type impellers, large paddle impellers, or segmented impellers are used to achieve nonturbulent bulk flow patterns at minimum shear rates. Turbine impellers as used in microbial bioreactors cause damage to many cell lines and sufficient mixing requires rather high stirrer speed. Pumping capacity can be maximized and mixing times minimized by the use of

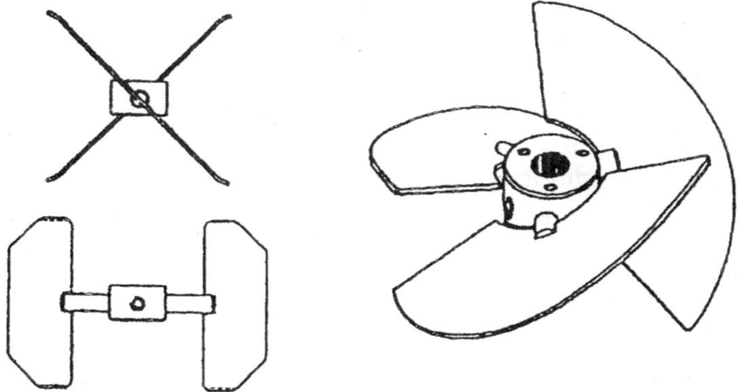

Fig. 6.3. Schematic drawing of large paddle impeller (left) and three-blade segment impeller (right).

large agitators operated at low stirrer rates rather than small impellers at high stirrer rates. As the reactor size increases, especially with increasing height, it becomes necessary to install multiple impellers to prevent compartmentalization of the fluid. In such setups attention has to be paid to optimum positioning of impellers regarding mixing times and homogeneity at a given stirrer speed. Many other impeller types have been suggested in the past. Anchor mixers provide superior mass transfer when using an aeration basket with silicon tubing wrapped around due to its radial transport characteristics. Feder and Tolbert suggested a sail impeller that should provide low shear homogeneous mixing. Comparative studies using a shear-sensitive inorganic test system revealed, however, that lowest shear forces as determined by disruption of the flocks were obtained with a three-blade segment impeller followed by a large pitch bladed impeller.

Different helical impeller types have been applied for insect and plant cell cultures. In combination with special baffles at the gas–liquid interface surface oxygen transfer rates of 4 up to 45 h^{-1} were reported for water and agar suspensions. The cell lift impeller is characterized by a very specific design for gentle mixing by pumping the fluid upwards through an axial cylinder and gas transfer in a surrounding aeration cavity covered by a mesh screen. The annular cage impeller is very similar in design but provides improved oxygen transfer. Even other mixer types such as the Vibromixer and the vibrating mixer described by Monahan are combined mixing and oxygen transfer devices. Blasey reported on the advantageous application of the Vibromixer for keeping BHK21 cells in single cell suspension. A special tumbling membrane stirrer developed by Lehmann (1–100 L bioreactor scale) provides gentle mixing and efficient gas transfer through hydrophobic polypropylene membranes. Despite the variety of impeller types described, marine impeller designs generating an upward fluid flow with a rather low radial dispersion are dominant for suspension culture in industrial bioreactor setups.

In the literature, a number of scale-up criteria valid for agitation rate and tolerable shear forces are discussed such as impeller tip speed, integrated shear factor, or Kolmogorov's theory of isotropic turbulence. For hybridoma cells a maximum tolerable tip speed of 1 m/sec is reported for laboratory scale (1–10 L) STRs. Other authors state tolerable tip speeds of up to 2m/sec. A very important consideration for cell damage by agitation is vortex formation with bubble entrainment since this phenomenon is very likely to cause cell death. Other authors have investigated the effect of agitation rate on cell growth and product formation with and without sparging. They found that increasing stirrer speed alone did not cause cell damage but in combination with direct sparging a decrease in cell growth rate and maximum density was observed. The use of baffles is avoided in animal cell culture to minimize the generation of shear forces. If the agitation rate is too low dead zones are formed causing cell sedimentation and subsequent irreversible aggregation.

Aeration

Typical aspect ratios (height–diameter) for stirred tank reactors vary between 1:1 and 3:1. At low aspect ratios gas transfer via the headspace of the vessel is improved due to the relatively high surface to volume ratio. Stirred tank reactors used for microcarrier culture are frequently designed accordingly. However, higher aspect ratios offer advantages when direct sparging is used for oxygenation of the cell culture. Better dispersion and longer residence times of gas bubbles in the culture liquid are obtained resulting in improved gas transfer rates. Direct sparging of air or oxygen is performed with ringspargers (drilled holes$>$0.5 mm diameter) similarly designed as those used in microbial fermenters or microspargers made of sintered stainless steel material (10–100 μM pore size). Especially in small bioreactors excessive sparging leads to considerable cell damage and death as the cells are disrupted when the bubbles burst at the gas–liquid interface. In larger bioreactors excessive sparging and foam formation can be minimized effectively by applying microsparging with pure oxygen at low flow rates with optimized pO_2 control parameters. The observation that shear stress through sparging is decreased as the reactor size increases is confirmed by Henzler's theoretical treatment of sparging in different reactor scales.

The gas–liquid interfacial phenomena and their effects on growth, viability, and productivity of animal cells have been extensively studied by many authors. Cell damage through bursting bubbles is even increased in serum-free or protein-free media due to the missing shear protection provided by serum or albumin. Therefore, surface active substances such as Pluronic F68, Polyvinylalcohol, Methocel, and other polymers are typically added to cell culture media. Furthermore, cells might be entrapped in stable foam layers generated by excessive aeration with microspargers. Foaming can be reduced by the addition of antifoam agents, although these agents may negatively interfere during downstream processing. Alternatively, the use of a hydrophobic net made of polysiloxane placed onto the liquid has been proposed. Despite the shear and foam related challenges, direct sparging is most frequently used in stirred tank reactors larger than 10 L due to the high oxygen transfer rates provided and the simple scale-up, handling, *in situ sterilization* (SIP) and *cleaning-in-place* (CIP) of the equipment.

Another problem encountered with microsparging is carbon dioxide (CO_2) accumulation as a result of the low flow rates of pure oxygen (<0.01 vvm) applied to minimize cell disruption and foam formation. Elevated pCO2 levels decrease the medium pH and may adversely affect productivity, cell growth, or glycosylation of protein products. Therefore, oxygen transfer and CO_2 stripping needs to be balanced carefully via optimized bubble size and airflow rate. Another possibility to remove excessive CO_2 is through intense ventilation of the headspace with air. Stripping can further be improved through a radial impeller directly below the liquid surface. Addition of HEPES (zwitterionic organic buffer) to increase the buffer capacity or sodium hydroxide or bicarbonate as corrective agents will, however, rather increase the pCO_2 and osmolarity and may even lead to cell damage due to high local pH values. To overcome these problems an optimized direct sparging strategy taking into account the CO_2 accumulation was recently proposed by a research group at Bayer.

Bubble-free oxygen transfer methods were suggested for stirred tanks via thin, gas permeable silicone tubes or microporous membranes made of PTFE or polypropylene. These systems have significantly lower oxygen transfer capacities when compared to direct sparging. Scale-up of membrane aeration devices is very limited and the largest vessels constructed provide 100–300 L working volume.

Arranging many meters of tubing to maximize mass transfer and mixing is an engineering challenge. In particular, sedimentation of cells, cell aggregates, or microcarriers may occur. Oxygen consumption rates determined for different cell lines (CHO, BHK, hybridoma cells, insect cells, hematopoietic cells) are in the range of 0.02–0.6 μmol/ 10^6cells/hr. Based on these data it can be calculated that an oxygen transfer coefficient k_la of 0.4–4 h^{-1} is necessary to supply sufficient oxygen to $\times 10^6$ cells per ml at a pO_2 of 30% air saturation. This rough estimation clearly indicates that at higher cell densities membrane oxygenation systems may not guarantee sufficient oxygen transfer even when using pure oxygen especially in a larger scale.

A detailed comparative discussion of oxygen transfer characteristics in stirred tanks using different oxygenation methods, i.e., surface, sparger, and membrane aeration, in connection with impeller design and medium additives such as Pluronic F 68 and albumin, is provided by Henzler, Aunins, and Moreira.

Perfusion

Suspension culture in STRs offers broad flexibility regarding operation mode. In the pharmaceutical industry typically batch, repeated batch and fed-batch processes are applied for production purposes. Continuous cultures such as in chemostat or cytostat mode have been mainly used for scientific investigation and optimization of cell growth, metabolism, and productivity under different conditions. During the last 10 years perfusion culture, i.e., the continuous exchange of spent medium while retaining the cells in the bioreactor has become an established technology. Perfusion culture has been used for production of monoclonal antibodies and recombinant proteins. Large-scale operations ($\leq$ m^3 bioreactor scale) are running at Bayer, Biovitrum, Chiron, and others. Devices for continuous harvest of cell-free supernatant from suspension cultures are based on either sedimentation or filtration as the separation principle. Critical parameters for the design of such devices are the size of suspension cells (10–30 μm), their low density [~5% greater than that of the medium], and their shear sensitivity. For filtration-based separation devices, the content of cell debris and macromolecules in the culture supernatant, i.e., supplemented or cellular proteins and released DNA, frequently cause problems in operation longevity due to membrane clogging.

During the last 20 years a large number of different devices for operation in an external loop or satellite vessel of the bioreactor or inside the vessel have been described in the literature. However, only a limited number of systems are commercially available to date. Generally, external systems can easily be exchanged upon failure and offer the advantage of scale-up via application of parallel units. However, controlling critical culture parameters such as pO_2, pH, and temperature is difficult and the need to use pumps to recycle the cell suspension may cause mechanical damage to sensitive cells. The most commonly used devices for suspension cell perfusion culture are continuous centrifuges, different types of settlers, external cross-flow filtration devices, and spinfilters made of metal screens, either operated in the bioreactor or an external loop.

In the following paragraph internal retention systems specific for stirred tanks will be described briefly. Microporous polypropylene membrane segments (0.2 μm pore size) of the above described membrane stirrer were hydrophilized with ethanol and applied for intermittent harvest of culture supernatant and feed of fresh medium or even recycling of amino acid enriched harvest. A static perfusion module equipped with permanently hydrophilic microporous PTFE membrane was used for hybridoma and insect cell culture in 2–10 L stirred tanks. Rotating wire cages, so-called *spinfilters*, mounted on the stirrer axis or driven via an independent motor, have been used with single cell suspensions (nominal pore sizes in the range of 5–20 μm) and aggregate forming cell lines (nominal pore sizes 40–120 μm). Several authors described continuous perfusion cultivation of suspension cells over weeks up to months at retention rates >90% leading to maximum cell densities far beyond 1×10^7/mL. Especially at low pore sizes frequent clogging has been reported. At larger pore sizes noncomplete cell retention is obtained for single cells (typical cell diameter 10–20 μm). Therefore, the pore size has to be carefully balanced with the desired retention rate to avoid early termination of the culture. Factors affecting the retention degree of suspension cells, filtration performance, and scale-up have been investigated by several authors but are still not completely understood. From the literature it can be deducted that pore size, surface velocity of the mesh, agitation rate, the fluid flow pattern in the stirred tank, and the type of applied mesh are important parameters determining spinfilter performance. Due to their mechanical robustness scale-up has been shown in stirred tanks up to several hundred liters. Despite the vast literature it is rather challenging to successfully design and operate spinfilter bioreactors for suspension cells due to the empirical nature of the data gathered so far.

A variety of sedimentation devices for suspension cells suitable mainly for small-scale cell culture have been developed. Settlers permit only slow perfusion rates for single cell suspensions since mammalian cells settle at a rate of 2–10 cm/hr and retention is therefore often not complete. Several authors applied systems based on gravitational settling, and Tokashiki succeeded to scale-up based on this principle to the 22 L scale and claimed a scale-up potential to some 200 L. Sato and more recently Hosoi perfused Namalva and hybridoma cells by applying centrifugal forces to facilitate settling in an enlarged hollow stirrer shaft. A major advantage is that failure due to clogging does not occur since separation is not carried out via a physical barrier and selective retention of viable cells can be achieved due to the two-fold lower sedimentation velocity of nonviable cells. This observation is generally valid for sedimentation-based perfusion devices and has been shown also for external centrifuges applied to recycle the cells back to the bioreactor and ultrasonic separation.

Airlift Reactors

In addition to STRs, airlift reactors are widely used for suspension cell culture. The largest published scale of airlifts applied for animal cell culture is 2 m^3, which is routinely used for monoclonal antibody production at LONZA. This relatively new bioreactor type has been scaled-up to 1500 m^3 for microbial fermentation and 17,000 m^3 for wastewater treatment demonstrating its enormous scale-up potential. There are several publications on the use of this bioreactor type for suspension growth of mammalian (hybridoma, BHK, CHO, Namalva) and insect cell lines. Even studies on microcarrier cell culture were reported although it has generally been considered impractical due to microcarrier aggregation at the air–medium interface. Airlifts can be regarded as a type of bubble column since mixing is provided by the introduction of gas bubbles at the base of a tall column. In bubble columns the ascending bubbles cause random mixing. In airlifts, fluid circulation is obtained by mechanical separation of a channel for gas/ liquid up-flow (riser) and a channel for down-flow (downcomer). These channels are connected at the top and the bottom of the column forming a closed loop. Fluid flow is driven by the density difference between the sparged fluid in the riser and the bubble-free fluid in the downcomer. Basically, two different geometric configurations can be distinguished: (i) external loop vessels where the circulation takes place through an external loop and (ii) baffled vessels with either a cylindrical draft tube or a simple baffle resulting in an internal loop circulation. In airlifts mixing and oxygen transfer are generally coupled, i.e., the gas flow rate is controlled by the oxygen demand of the culture and determines the hydrodynamic conditions in the bioreactor. In order to decouple mixing and oxygen transfer, pH and pO_2 can be controlled by varying the composition cf a carrier gas (air or nitrogen) regarding oxygen and CO_2 at a maintained total gas flow rate. Typical superficial gas velocities are in the range of 0.001–0.01m/s and corresponding $k_l a$ values of 0.7–20 h^{-1} were reported.

Major design considerations are geometrical configuration of the riser and downcomer, aspect ratio, sparger layout, positions for electrodes, sample ports, feed and base addition. The most widely used design for animal cell culture is a bubble column with a concentrical draft tube. Typical aspect ratios are in the range of 6:1

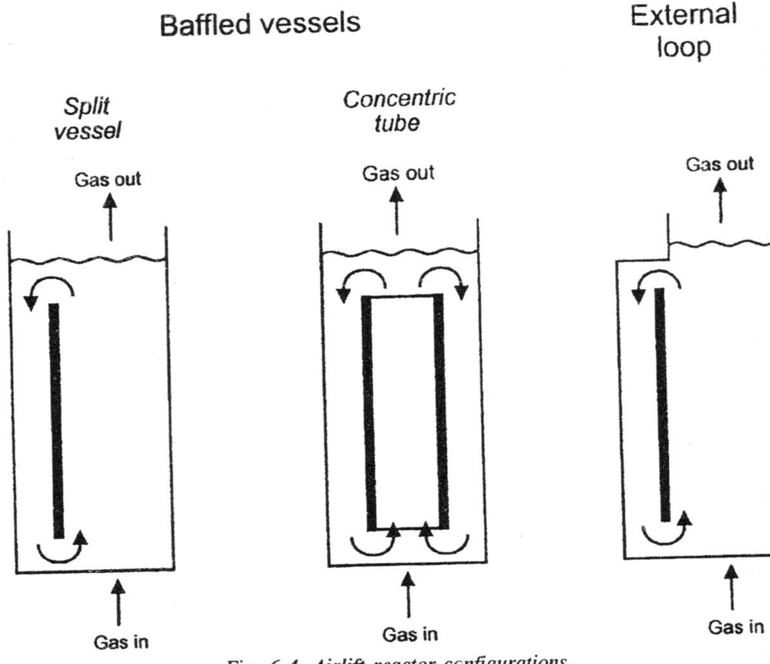

Fig. 6.4. Airlift reactor configurations.

to 12:1. Optimum mixing is obtained by keeping the cross-sectional area of the downcomer similar to that of the riser. It has also been shown that the cylindrical area below and above the draft tube is affecting mixing.

The main advantage quoted for airlift over STRs beside its ease of scale-up is that no moving parts and mechanical seals are needed which improves the reliability of sterile operation. Despite these advantages the airlift is not as widely used and modified for cell culture as the STR. This has historical reasons but might also be due to the limited flexibility in terms of working volumes and suitability for microcarrier culture. Another advantage frequently quoted is its gentler mixing action and suitability for shear-sensitive cells. However, similar considerations as described for direct sparging in stirred tank reactors are valid for cell damage and CO_2 accumulation.

A number of authors have studied the shear sensitivity of different cell lines such as hybridomas, BHK and insect cells in airlift and bubble column bioreactors and the effect of serum or surfactant concentration, sparger design, bubble size, and column height on cell survival. Although most reports on the use of airlift bioreactors are based on batch culture, it is generally possible to operate the system at high cell densities using external cell retention systems similar to those described for STRs. For example, Hulscher used an external settler for the selective recycle of viable cells to an airlift loop reactor.

Modifications

A modification of the bubble column principle is the inclined plate bioreactor designed to reduce the effect of bursting bubbles on cells. The inclination of the column leaves the bulk of the liquid bubble free while still providing effective liquid circulation. Another design modification to minimize cell damage due to sparging is provided in the bubble bed bioreactor. In this stirred tank-bubble column hybrid bioreactor the residence time of bubbles was dramatically prolonged by floating the bubbles in a countercurrent flow produced by an impeller in the lower part of a central conical draft tube.

Other Reactor Types

Hollow fiber bioreactor

In contrast to the homogeneous bioreactor systems described so far, which support in the range of 10^7 cells/mL under perfusion conditions, the heterogeneous *hollow fiber bioreactor* (HFBR) is a high-intensity system where tissue-like cell densities ($>10^8$ mL^{-1}) and structures are obtained. This bioreactor type was originally developed by Knazek in 1972 starting from commercially available ultrafiltration modules containing cellulose acetate capillaries and further developed by adding gas permeable silicone polycarbonate capillaries for improved oxygen supply and CO_2 removal. The units are composed of a large number of semipermeable capillaries potted into a cylindrical housing. In many reports especially from research laboratories simple hemodialysis cartridges have been applied. However, during the last 30 years a number of companies have commercially developed and automated systems for mammalian cell culture.

The membranes are either of an ultrafiltration type (10–100 kDa) or microporous (0.1–0.2 μm). Different membrane materials, for example, cellulose acetate, polypropylene, and polysulfone (0.2 μm), have been used. In most applications the cells are grown in the *extracapillary space* (EC) while oxygen enriched medium is recirculated through the fibers (*intracapillary space*, IC). In many cases, 10 kDa ultrafiltration membranes are applied to enable the use of inexpensive basal medium where expensive large molecular weight growth factors are supplied only in the EC space. Cell secreted proteins are also

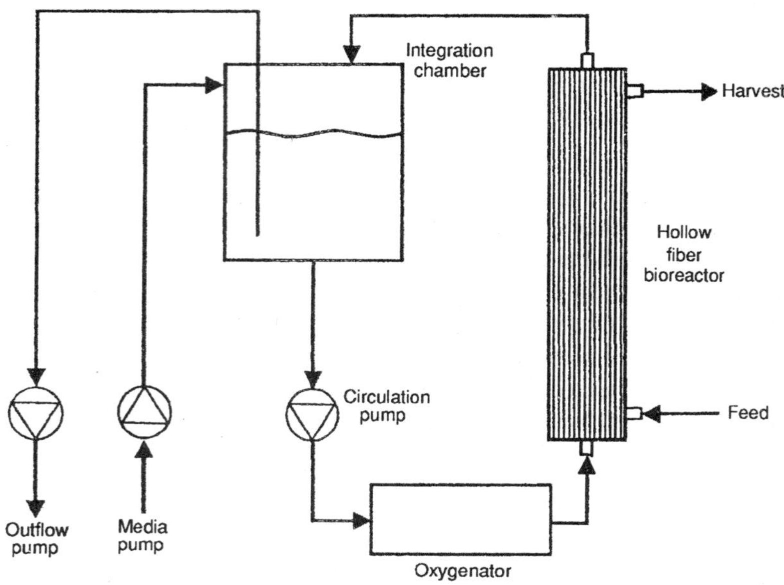

Fig. 6.5. Schematic drawing of a hollow fiber cell culture bioreactor.

retained in the EC space leading to a significant accumulation of the product in the EC space which substantially reduces downstream processing efforts. However, it was reported that inhibitory components might accumulate in the EC space, which negatively affect cell growth and productivity. Furthermore, proteases secreted by the cells or released from dead cells may cause product degradation during long-term operation. Due to product accumulation this bioreactor type is not suitable for toxic or feedback inhibited products. To improve nutrient supply and waste removal, protocols have been developed where fresh medium is supplied to a medium reservoir connected to the IC space after removal of spent medium. In other cases the medium reservoir is regularly replaced with fresh medium.

The basic concept of an axial flow hollow fiber bioreactor has been developed further over the years and a number of alternative designs have been described. An overview of cell culture systems and related operation modes and transport phenomena was given by Tharakan et al., Piret and Cooney, and Brotherton and Chau. Uludag et al. reviewed the technology in relation to its application in cell therapy and tissue transplantation.

In principle, three operation modes of hollow fiber systems can be distinguished: (a) open shell ultrafiltration, (b) closed shell

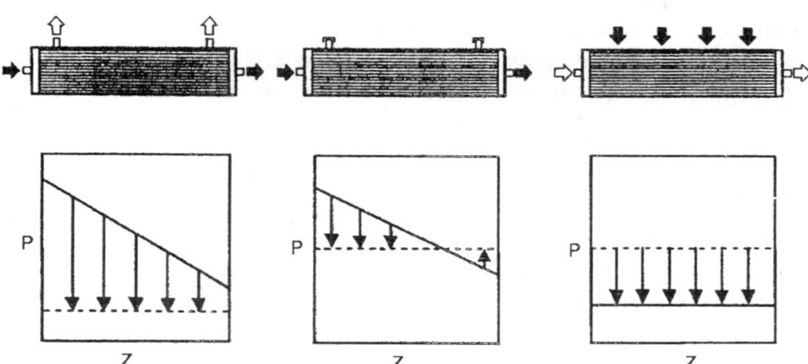

Fig. 6.6. Three operation modes of hollow fiber system (top) and their corresponding transmembrane pressure profile P (bottom).

ultrafiltration, and (c) crossflow ultrafiltration. In the open shell ultrafiltration mode the transmembrane pressure is decreasing along the length of the fibers. In the closed shell ultrafiltration mode the transmembrane pressure is positive in the entrance half of the module and a back-flow of medium from the EC is obtained in the distal part. These pressure gradients along the axial flow bioreactor types (a, b) cause nutrient and oxygen gradients, which may result in uneven cell growth. These gradient phenomena are the reason for the limited scale-up potential of these modules and a solution is to use multiple parallel units.

Attempts to overcome the gradient-related problems have been made by modifying the operation of conventional HFBR. Manipulation of the bioreactor orientation or the osmotic environment of the EC space were suggested in addition to periodically reversing the flow direction in the IC space or change the pressure in the EC space to generate a backflow of medium into the IC space. In order to improve oxygen supply to the EC space oxygen carriers were added to the medium that was recirculated through the IC space, which led to improved antibody production. Another approach is the development of alternative designs. In bioreactors with extra capillaries for oxygen supply the IC flow rate can be reduced and axial oxygen gradients are eliminated. By additional rotation of the culture chamber further improvement of the culture performance was claimed. Cima et al. developed a radial flow hollow fiber reactor based on concentric microporous polypropylene fibers with a 200 μm annular space for cell growth. They investigated different operation modes resulting in a more uniform distribution of nutrients by forcing medium through the EC space. The modeling of an intercalated dead end hollow fiber bioreactor with arterial (inlet) and venous (outlet) fibers is presented. An example for the crossflow operation mode is the flat-bed bioreactor developed by Ku et al. where gradients are substantially reduced compared to the axial configurations. The fibers are used for oxygen supply and pH control. The cells were retained in the unit using microporous stainless-steel filters. However, in most commercially available systems modified axial flow configurations are used.

Hollow fiber bioreactors have mainly been used for the growth of suspension cells. However, using other membrane materials such as polypropylene tissue-like cell densities of adherent cells were obtained and could be maintained over several weeks to months. Also coating of the fibers with polycationic materials such as polylysine can be used to enable adherent cell growth. Many different types of mammalian cells have been cultured in hollow fiber bioreactors such as primary cells, tumor cells, and stable cell lines. The most important application is the production of monoclonal antibodies mainly for diagnostic and research purposes. Considerable attempts have been made to use the technology for the production of therapeutic antibodies, which resulted in the first registration of a hollow fiber produced drug. Other, more recent applications are the production of cells for cell and tissue therapy and the use as artificial organs (liver, kidney, pancreas).

Miscellaneous dialysis bioreactors

Several other bioreactors based on the dialysis principle were developed in the past.

Generally, dialysis culture offers the advantage of continuous supply of nutrients and removal of low molecular weight metabolic waste products while accumulating the product in the cell culture compartment. The stirred tank dialysis bioreactor setup eliminates the formation of gradients, the major disadvantage of the hollow fiber bioreactor. Compared to stirred-tank perfusion systems mass transfer is still diffusion controlled. Culturing hybridoma cells in simple dialysis bags placed in a spinner was further developed to setups where dialysis membranes or dialysis modules were submerged in STRs. In the latter systems the cells are cultured in suspension in the stirred vessels and medium from a reservoir is recycled through the dialysis membranes. Cell densities of more than 1×10^7 mL^{-1} and significantly increased product concentrations have been reported in an industrial application. Another modification of this principle is a two-compartment stirred tank vessel with cells on one side of the dialysis membrane and dialysis medium on the other side. Dialysis bioreactor technology was reviewed by Portner. Some historical examples of complex systems mainly for cell maintenance are the InVitron *Static Maintenance Reactor* or the *Membroferm bioreactor*.

Microencapsulation technique

A completely different technology for suspension cell culture based on immobilization and compartmentalization is microencapsulation. It was originally introduced by Lim and Sun for the immobilization of mammalian cells used in bioartificial organ applications. This technique has been further developed and applied to monoclonal antibody production for commercial use. The cells are suspended in a solution of a naturally gelling polymer and the microspheres are generated subsequently. The microcapsules can be "*cultured*" in suspension in different bioreactor types. Various materials have been described for suspension cell entrapment in microcapsules; Ca-alginate, Na-alginate, agarose, cellulose sulphate. Additionally, composite gels combining the properties of alginate and agarose or PEG and alginate were developed. Collagen and fibrin were suggested for the cultivation of anchorage dependent cells. The diameters of the capsules range from 0.5 to 1 mm. Agarose beads have a lower mechanical strength compared to alginate. Alginate beads have been coated with polylysine to generate a semipermeable membrane with controllable molecular weight cut-off. Improvements of the mechanical strength were obtained using photosensitive polymers. The cell leakage of conventional alginate could be reduced using PEG–alginate composite beads. Even more materials and modications are used for cell and gene therapy and tissue transplantation as reviewed by Uludag.

Microencapsulation of cells is a well-established technology in bacterial culture and especially in wastewater treatment. Much of the literature on the preparation of the microcapsules, their rheological and mechanical properties, as well as transport and mass transfer phenomena is found there.

Typical advantages of microencapsulation are the shear protection of cells, the product concentration, and compartmentalization obtained with some microencapsulation methods [e.g., polylysin coated Na-alginate beads] and the high cell densities in the particles in the range of 10^7 mL^{-1} to 10^8 mL^{-1}. Another advantage of the microencapsulation technique often cited is the increased specific antibody production rate. However, a critical review of the literature shows that an increase in specific antibody productivity is cell line specific and dependent on the cultivation conditions and applied parameters. Furthermore, serum-free and even protein-free cultivation could be established facilitated by the compartmentalization of cells and the bulk of the medium. However, a drawback of the technology is the diffusion-controlled transport of nutrients, waste products and—most critically—oxygen, which may lead to necrosis in the center of larger microcapsules. In case of product retention in the capsule this system is not suitable for proteolytically sensitive and feedback inhibited products.

Microcapsules have been used in different bioreactor types such as stirred tanks, airlifts, and fluidized beds mainly in small-scale batch operation mode. Although, from a historical perspective, the development of this technology was stimulated by the need for efficient and economic production methods for monoclonal antibodies it cannot be considered as suitable for industrial-scale mammalian cell culture. The major application area of this technology today is cell and gene therapy and tissue transplantation.

Bioreactors for Anchorage-Dependent Cell Cultures

Small-Scale Culture Systems

Adherent cell culture has a tradition going back to the beginning of the last century when tissue culture was first devised as a method to study the behavior of animal cells free of systemic variations. Basic techniques are described by Freshney and an interesting historical review on the advances in tissue culture during the last century is given by Jensen. In anchorage-dependent monolayer culture, the achievable cell number is directly proportional to the available growth surface. Cell yields in the order of 10^5 cells/cm^2 are obtained for HeLa cells, CHO and HEK 293 cell lines. Typical small-scale culture systems include Petri dishes and multiwell pates that are kept in a humidified CO_2 incubator. From a sterility point of view T-flasks available with and without vented caps are preferable (available surface area 25–225 cm^2, 500 cm^2 triple layer. Culture flasks used to be made of glass. However, nowadays, disposable plastic ware is standard. Coating the plastic surface with l-lysine or other substances may improve attachment of less anchorage dependent cells such as HEK 293 cells. Especially in roller bottle culture this helps preventing the formation of suspended aggregates. One of the simplest systems for scaling up monolayer cultures is the Cell Factory providing culture surface areas of 600–24,000 cm^2 and larger (100,000 cm^2) via multiple interconnected plastic layers. Gas transfer is obtained via diffusion and a critical liquid depth of a few millimeters should not be exceeded. Spier reported oxygen transfer rates of 0.53 μM/cm2/hr for monolayer aeration.

Especially for commercial vaccine production roller bottles have been applied for over 40 years(surface area 850 cm^2, 1500 cm^2, and other sizes). Roller bottles require a specific apparatus for constant rotation. Major advantages of this traditional "*large-scale*" culture method are (a) the increased surface area, (b) the constant mixing preventing gradients, (c) very high oxygen transfer rates, and (d) the flexibility in number of units applied for a certain task. Handling large numbers of multiple culture units for industrial production of vaccines or recombinant protein is very tedious and prone to contamination. Therefore, successful attempts have been made toward automation using laboratory robots such as the Cellmate, which is able to handle roller bottles and T-flasks.

The CellCube system (Corning Costar, available surface area 8500– – 85,000 cm^2) combines the advantage of a large surface area typical for multilayer systems with good oxygen transfer and mixing capabilities. Medium is continuously pumped through the system which can be oxygen and nutrient enriched in an external loop This system has been used at Merck for the development of Hepatitis A vaccine and by other authors for virus propagation.

Although the described methods are feasible for large-scale production there are considerable drawbacks such as (a) high costs for labor, equipment, and consumables, (b) considerable risk of contamination, and (c) relatively poor opportunity to control culture parameters at optimum set-points. This led to the development of scaleable bioreactor systems for anchorage-dependent cell culture. The most prominent examples are microcarrier culture in stirred tanks, fluidized, or packed beds or cell culture in solid bed bioreactors. The initial impetus for the development of such techniques was provided by the mass vaccination campaigns of the 1950s against viral diseases and later by the advent of recombinant DNA technology used for the production of drugs in animal cells.

Microcarrier Culture in Stirred Tank Reactors

The basic idea of microcarrier culture was to develop a unit operation system with similar scale-up and environmental control potential as suspension cell culture in stirred tanks. Van Wezel introduced the use of dextran carriers for the growth of human fibroblast like cells in 1967. A historical review of the development of microcarrier technology is presented. In principle, this technique comprises the cultivation of anchorage dependent cells on small solid particles suspended in growth medium. Under proper conditions the cells attach and spread on the carrier surface and grow to a confluent monolayer. A series of different types have been introduced and attention has been paid to the chemical and physical properties of the carriers such as size, density, and surface charge. General requirements are nontoxicity, good adhesion properties, and a buoyant density of 1.03–1.04 g/mL to be easily kept in suspension. Furthermore, uniform size distribution, suitability for microscopic monitoring of cell growth, autoclavability, and a high batch-to-batch consistency are important quality features for industrial application of the technology. Typically, the carrriers have diameters of around 100–200 μm and are made of dextran, plastic, gelatin, glass, or cellulose. Even the use of flat microcarriers in STRs has been reported. For good cell attachment a number of physicochemical factors are critical; a positive or negative surface charge, the charge density, the presence of cations in the medium. Very often the surface is coated or modified in a specific way to improve cell attachment. In this regard, essential proteins of the extracellular matrix such as collagen, proteoglycans, fibronectin, laminin, elastin, and chondronectin affect cell adhesion positively. A comparative study of cell growth and product formation for 13 different carrier types using a recombinant CHO cell line producing monocyte-colony inhibition factor showed clear differences in cell growth and specific productivity for the different carrier types.

In 1986, porous microcarriers made of gelatin were introduced and have been used in many applications since than. Due to the porous structure an increased surface area in comparison to smooth beads is obtained, which enables to achieve an improved maximum cell density and productivity. This carrier type is also suitable for the propagation of suspension cells that can be entrapped in the pores. Furthermore, it was claimed that shear sensitive cells are protected inside the beads. Other materials were used to produce porous carriers for application in STR such as silicone and cellulose. Probably the commercially most important microcarrier types for stirred tank culture are Cytodex and Cultisphere carriers.

This technology has primarily been used for the large-scale production of vaccines and interferon up to the several thousand liters scale and also for the production of recombinant therapeutic proteins . For these applications, harvesting and separation of the cells from the carriers is not necessary except during volume expansion when cells need to be detached to inoculate a larger bioreactor. Attempts made to use bead-to-bead cell transfer have never been very successful despite for certain cell lines under specific conditions. Therefore, in general, methods are used where the cells are detached enzymatically or by using buffer solutions containing chelating agents such as EDTA. In case of the gelatin beads enzymatic treatment leads to the complete dissolution of the carriers, which is facilitating subsequent manipulations such as cell harvest and inoculation. Other applications of stirred tank microcarrier culture are the production of proteins and cells used in drug discovery and research.

For adherent cell culture in stirred tanks suggested microcarrier densities range from 1 g/L under batch up to 10 g/L under fed-batch or perfusion conditions providing cell densities of 10^6 mL^{-1} to more than $10^7 mL^{-1}$. In contrast to suspension cell culture in STRs retention of immobilized cells on the relatively large carriers is easily and reliably achieved with spinfilters (pore sizes of >75 μm) or even simpler means such as stagnant zones. Problems observed during perfusion of single cell suspension such as filter fouling or incomplete cell retention have rarely been reported.

Attention has to be paid to mixing and aeration of microcarrier cultures in stirred tank reactors. Especially cells grown on smooth carriers are sensitive to overagitation. Growth arrest and even detachment of cells has been observed above critical stirrer rates. The stirring regimen is even more important during inoculation where intermittent profiles have been suggested to improve attachment. Several researchers investigated hydrodynamic effects on cells in agitated carrier cultures. Three potential mechanisms of cell damage were postulated: bead collision with the impeller or other stationary surfaces, bead-to-bead collision, and bead interaction with turbulent fluid eddies of the same size as the beads. Gregoriades found that cell damage of cells attached to microcarriers was observed at energy dissipation levels orders of magnitude lower than those values reported to cause damage to suspended animal cells. However, hydrodynamic stress can be controlled at tolerable rates providing a well-mixed culture by choosing large marinetype impellers that generate a strong axial flow at low stirrer rates.

Another problem observed during scale-up is oxygen supply. Generally, bubble-free aeration methods as described earlier in this chapter are preferable for carrier culture since microcarriers tend to be entrapped in foam layers generated by direct sparging and cell damage has been observed. Through the optimization of direct sparging (control algorithms, gas mixture, pore size of the sparger, gas flow rates) foam formation can be minimized and this oxygenation method becomes applicable for carrier culture especially in larger scale. In contrast to freely suspended cells, it has been shown that cells attached to microcarriers are not damaged by bubble rupture at gas-medium interfaces. Instead it was suggested that cell damage is caused by bubble-attached cells being removed from the bead as a consequence of the hydrodynamic drag force of the rising bubble. In the same study different surfactants were tested for their suitability to protect the carrier attached cells but none gave satisfactory results.

In conclusion, microcarrier culture in stirred tank reactors is a well-established, scaleable technology for anchorage-dependent cells offering a series of advantages such as high growth surface-to-culture volume ratio, simple cell separation, perfusion, and direct monitoring and control. Homogeneous microcarrier culture was also reported in airlift bioreactors although it is not as established as stirred tank cultivation.

Fluidized Bed Bioreactors

The fluidized bed technology was introduced for animal cell culture in the 1980s by the company Verax Corporation. The cells were grown on porous microspheres, made of collagen and weighted with a noncytotoxic steel to achieve a specific gravity of >1.6, so that the carriers remained suspended in the high-velocity (~70 cm/min) upward-fluid flow of culture medium. The microspheres (500 μm diameter) had a sponge-like structure of interconnected pores and channels with a diameter in the order of 20–40 μm allowing the cells to enter easily and populate the interior of the carrier. The fluidized bed was contained in a column type bioreactor that was connected to an external recirculation loop. A gas exchanger (hollow fiber cartridge), pO_2, pH, and temperature sensors, heating elements, and a circulation pump that controlled the expansion of the fluidized bed were located in the loop. Dissolved oxygen levels were monitored at the inlet and outlet of the gas exchanger in order to measure the oxygen transfer rate of the reactor and to control the oxygen flow. Typically, oxygen transfer rates approached 10 mmol/L/hr. Carbon dioxide could also be supplied to the gas exchanger for the purpose of pH control. Should the cell culture medium become to acidic addition of base kept the pH at a given set point. This type of bioreactor, now no longer on the market, was available from 0.4 L (research scale) to 24 L expanded bed volume size for production of proteins for human use. The design principles of the bioreactor as well as examples for cultivation of hybridoma and

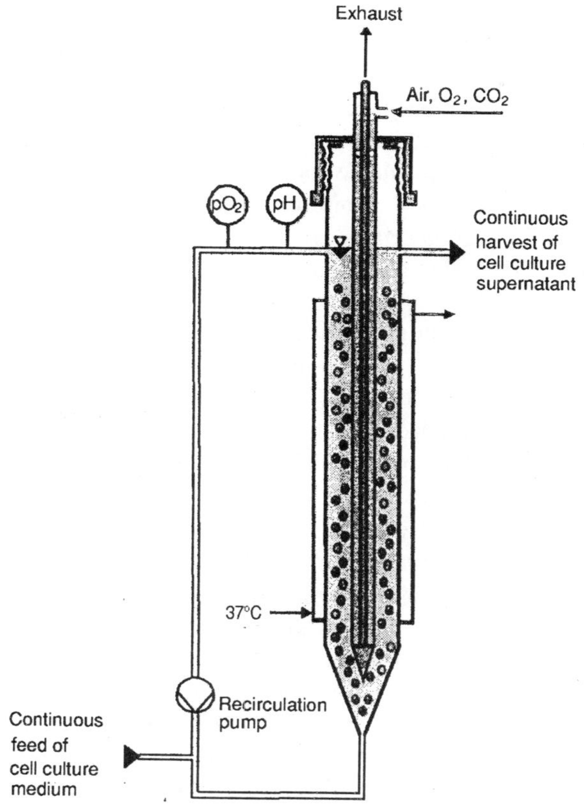

Fig. 6.7. Schematic drawing of the fluidized bed bioreactor system with in-line aeration module for bubble and gradient free oxygen supply.

recombinant CHO cells at different scales are comprehensively. The general concept of the Verax-system inspired many research groups to investigate and optimize the fluidized bed technology for animal cell culture.

One focus of research was the microcarrier itself. The growth of cells on different matrix materials, such as glass and polyethylene, was investigated. Porous Siran carriers made of borosilicate glass were shown to be a cost-effective alternative to the weighted Verax microspheres. Chemical modifications of borosilicate Siran glass resulted in similar cell densities per milliliter packed bed volume as reported for the Verax microspheres. In contrast to stirred tank systems, a significantly higher specific gravity of >1.5 of the carriers is typically chosen to achieve homogenous fluidization in a fluidized bed bioreactor.

A major disadvantage of the fluidized beds was the progressive depletion of oxygen from the bottom to the top of the expanded bed. This problem of an oxygen gradient along the axis of the tubular reactor could be overcome by integration of a membrane oxygenation module directly into the fluidized bed. Alternatively, the integration of an in-line gasification tube module developed at the Research Center Julich was suggested. This latter fluidized bed bioreactor with improved oxygenation is commercially available form B. Braun Biotech International. This bioreactor type is available with settled carrier volumes ranging from 0.02 to 0.5 L. Typically, it is operated with gelatinized porous Siran microcarriers that are available in different diameters. This system is used in the biotech industry for large-scale production of proteins for research and medical applications.

An alternative fluidized bed system based on internal recirculation of the culture medium was developed at the Institute of Applied Microbiology in cooperation with Vogelbusch GmbH, Vienna, Austria. This system is available from laboratory scale to production scale.

The *Cytopilot bioreactor* comprises lower and upper cylindrical chambers. A draft tube in the fermentor replaces the external recycle loop. The lower chamber houses an axial flow impeller that provides a fluid flow to expand the bed. Oxygen is homogeneously microsparged into the downcomer close to the impeller and uniformly distributed in the upper chamber by a designed liquid flow and a gas distribution plate at the bottom entrance of the fluidized bed. The bed expands or contracts as a function of the stirrer speed that creates the necessary hydrodynamic pressure to lift the settled microcarriers. The lower chamber is additionally equipped with a heating circuit (double water jacket), sampling and harvest ports, pH and pO_2 sensors.

Bluml and colleagues developed a microcarrier type that consists of polyethylene weighted with chalk and silicates. The buoyant density of this carrier type is dependent on the ratio of these three compounds providing various densities suitable for use in stirred tank, fluidized bed or packed bed systems. This former "*IAM-carrier*" is now commercialized as Cytoline and is very often used in combination with the Cytopilot.

The application of different types of microcarriers is described elsewhere in this book and has been reviewed previously. Verax microspheres, Siran and Cytoline carriers have been successfully evaluated for protein production with anchorage dependent cell lines and suspension cells such as hybridomas. Very often "*laboratory home-*

made" bioreactor designs were used for all kinds of comparative studies.

Reliable determination of the cell density in immobilized cultures used to be a draw-back when using Siran or Cytoline carriers, since both carrier types cannot be solubilized enzymatically as is possible with proteinaceous materials such as collagen. This problem of growth monitoring has been overcome recently by the use of dielectric spectroscopy.

Packed Bed Bioreactors

Fixed bed or packed beds are an alternative bioreactor technology to fluidized beds for the immobilization of cells. Both terms are used as synonyms for immobilization systems that have been deduced from large-scale fixed bed reactors originally designed for wastewater treatment. Initial work with mammalian cells started in the 1950s and aimed for tissue-like growth of cells in small-scale packed beds of cellophane, Raschig rings, and glass spirals. However, glass beads became the most widely used matrix. In 1976, Spier and Whiteside reported the production of foot and mouth disease vaccine in a packed bed system that could be scaled up to a unit process containing 100 L of medium. Glass-bead systems were considered for the production of interferon, herpes simplex virus, tissue plasminogen activator, and acetylcholinesterase.

The original design of packed beds was very simple and required two connected vessels; one holding the bed material, a second containing the culture medium and instrumentation for maintaining temperature, pH, and dissolved oxygen. The packed or fixed bed systems, initially operated with solid glass beads, did not achieve widespread use, mainly because they had a low surface area per unit volume and they were not suitable for suspension cells. These disadvantages were overcome by replacement of solid glass beads through porous glass spheres. Porous carriers were soon considered to have a greater potential for scale-up regarding both achievable cell density and increased surface-to-volume ratio.

Typically, packed bed systems are composed of a cylindrical bioreactor chamber filled with porous glass beads, a gas-exchanger (hollow fiber module or a stirred tank with air/oxygen sparging), a medium reservoir, and a pump that circulates medium between the bioreactor and the reservoir. As an alternative to glass carriers, other macroporous materials have been placed into bioreactors to support cell growth. The packed bed culture systems are designed for the immobilization of anchorage dependent or suspension cells providing high cell densities in the macroporous matrixes for the production of secretory proteins and lytic viruses. Some of the major advantages of packed or fixed bed reactors are the low surface shear rates, the high cell density, and productivity achievable and that no particle–particle abrasion is obtained. Reported disadvantages are the blockage of the pores due to high cell densities, poor oxygen transfer and the risk for medium channeling in the bed. Porous Siran carriers of 3–5 mm diameter were shown to give highest cell yields. This is due to the open bed structure (~ 1 cm^2 channel cross-sectional area) and reduced channel blockage caused by increases in biomass, uneven distribution of the inoculum and media channeling within the bed.

Alternative matrices such as nonwoven fibrous polyester disks were evaluated and packed into column-type reactors or in a basket mounted into a modified stirred tank reactor. A schematic drawing of the CelliGen Plus bioreactor. These two alternative setups were used for the production of recombinant proteins and viruses. A similar fermentor concept has been developed by Meredos GmbH, Bovenden, Germany.

Packed or fixed bed perfusion bioreactors have been investigated for other applications than protein production. Research teams evaluated collagen coated reticulated polyvinyl formal resin (PVF) as a matrix for primary hepatocyte cultures for use as a bioartifical liver. Other researchers investigated the *ex vivo* expansion of hematopoietic cells in porous glass carriers for bone marrow transplantation or gene therapy applications. Packed bed bioreactors mimicking lung and liver function were characterized as potential model systems for pharmacokinetic analysis of new drugs.

Other Reactor Types

The Opticell system, an automatically controlled system for anchorage-dependent and suspension cell culture, was developed in the 1980s. Anchorage-dependent cells were grown on a nonmodified ceramic matrix whereas suspension cells required a modified ceramic matrix. The scalability of this system is proportional to the amount of surface provided and is limited in contrast to stirred microcarrier systems, fluidized bed and packed bed bioreactors

A dialysis bioreactor with a radial-flow fixed bed for animal cell culture was derived form a dialysis membrane bioreactor used for microbial fermentation. This bioreactor consisted of two chambers, which were separated by a cylindrical membrane. The inner chamber was equipped with a centrifugal pump and packed with porous glass spheres and the centrifugal pump induced the radial flow of the medium. Aeration was done in the outer chamber that also served as medium reservoir. The fiber bed bioreactor was designed for anchorage dependent cell culture and was composed of a concentric-cylinder airlift reactor, in which the annulus is a packed bed of glass fibers. In this bioreactor, oxygen containing gas is sparged into the inner draft tube. Bubble-free medium flows down through the fiber bed in the outer cylinder providing convective oxygen and nutrient transfer to the cells. This bioreactor concept has been applied to large-scale production of murine bone marrow cells. Design and scale-up of a bioreactor based on cell attachment and growth on Koch-Sulzer static mixing elements was reported for viral vaccine production.

Bioreactor Operation Modes

The following operation modes for cell culture bioreactors have been described in the literature; batch, fed-batch, continuous culture without cell retention (*chemostat* or *cytostat*), and continuous culture with cell retention (*perfusion*). During perfusion culture sometimes a cell bleed is applied or obtained due to noncomplete cell retention. The kinetics that is observed during this type of continuous culture is a combination of the two types of continuous mode, i.e., perfusion and chemostat. The different modes are discussed in more detail in this chapter. Common for the different operation modes is that the temperature, the oxygen partial pressure, and the pH of the culture are controlled. This means continuous supply of oxygen and removal of carbon dioxide.

Batch and Repeated Batch Cultures

In batch culture the bioreactor is charged with cells and medium, no further medium is added or withdrawn during the process and at the end of the culture the whole reactor content is harvested at once. The cells grow exponentially while nutrients are consumed and metabolic waste products are accumulated until the maximum cell density for the given medium is reached due to nutrient limitation or waste product inhibition. Historically, ammonium, lactate, and CO_2 have been considered as inhibitory metabolic waste products. However, there are reports in the literature on small inhibitory proteins suggesting a key role in growth control. The length of the subsequent stationary phase is dependent on the cell type and finally the cells are dying during the declining phase. Sometimes a short lag phase after inoculation is observed which has been attributed to conditioning of the medium or the age of the preculture. In standard cell culture media cell densities

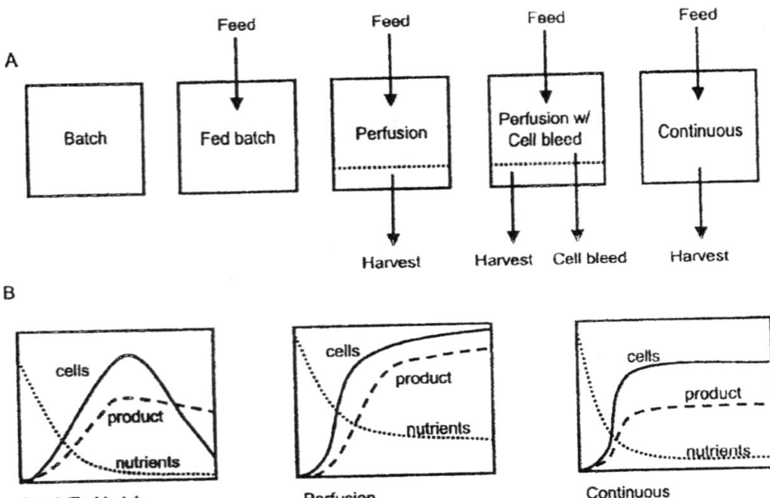

Fig. 6.8. Operation principle (A) and kinetics (B) of cell growth, nutrient consumption, and product formation during batch, fed-batch, perfusion, and continuous operations.

in the range of $1-3\times10^6\ mL^{-1}$ are reached with continuous cell lines. Significantly higher densities of $\geq10^7\ mL^{-1}$ have been reported for hybridoma cells using fortified media. However, this may not be generally achievable due to growth inhibition by metabolic waste products. Product formation can be growth associated, i.e., the specific productivity increases with increasing growth rate and ceases when the culture is entering stationary phase, or nongrowth associated, i.e., product formation increases with decreasing growth rate or is only observed during stationary phase. Several authors published data analyzing product formation in relation to the culture phase. The product formation kinetics has tremendous impact on the time of harvest during batch culture and the choice of cultivation mode in general.

Batch cultivation is simple and reliable and therefore the method of choice for many industrial applications. However, there has been an on-going debate on the pros and cons of continuous versus batch culture. Due to its simplicity it is also the most frequently applied mode for investigations of medium and culture parameters on growth, product formation and metabolism in process development, optimization, and scale-up. Therefore, most of the data available for commonly applied cell lines are derived from batch culture.

Repeated batch culture is very similar to batch culture with the exception that a fraction of the cell suspension is left in the bioreactor

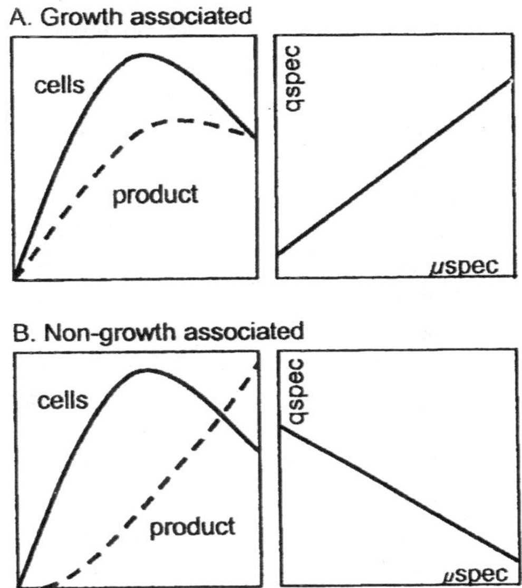

Fig. 6.9. Basic types of product formation kinetics during batch operation.

when the culture is harvested and fresh medium is added for a new batch growth cycle. For the success of this culture mode a high viability of the remaining cell suspension is important and optimization of the time when to initiate a new batch is necessary.

Fed-Batch Culture

During fed-batch culture a continuous or intermittent feed of nutrients is applied to prolong cell growth and product formation in comparison to simple batch operation. Since no cell suspension or supernatant is withdrawn the bioreactor volume increases until the whole content is harvested at once. In fed-batch operation, in general, higher product concentrations are obtained due to the prolonged accumulation of product. Controlled feeding of key nutrients is often performed and the nutrient concentration in the culture is maintained at low levels, aiming to minimize the formation of toxic metabolites, e.g., lactate and ammonia to improve productivity. These approaches mainly involve development of nutrient feeding solutions and feeding rate control strategies to avoid nutrient depletion and to reduce lactate and ammonia formation by controlling glucose and glutamine concentrations at low levels.

This cultivation mode has been applied widely to hybridoma culture and insect cell culture. Less reports are found on CHO fedbatch culture, however, it is a well-established technology for large-scale production of recombinant protein and antibody medicines. It has been applied to a range of cell lines such as HEK 293, BHK, FS-4. Using fed-batch mode the culture duration is typically prolonged from a few days to weeks. The cell density and productivity increase is dependent on the cell type. In insect cell fedbatch culture densities in the range of 5×10^7 mL^{-1} have been reported, which is equivalent to a 10-fold increase over batch. In CHO fed-batch cultures cell densities in the range of 5×10^6 mL^{-1} have been reported. In hybridoma culture the density could be increased to around $3–17 \times 10^6$ mL^{-1} and productivity up to 10-fold compared to simple batch operation. Using fed-batch culture in combination with nutrient-fortified medium even higher cell densities ($> 10^7$ mL^{-1}) were reported. The mathematical treatment was described by Glacken. In conclusion, fed-batch culture mode has been applied to improve productivity over conventional batch mode. Although it is a more sophisticated strategy with regard to process control it is still equipment-wise sharing the advantages of the simple and robust batch culture technique.

Continuous Culture

Continuous culture is characterized by a continuous flow of fresh medium into the bioreactor and removal of cell suspension from the culture system at the same rate, keeping the reactor volume at a constant level. An important criterion is that the reactor content is well mixed to ensure that the concentrations of cells, product and metabolites in the reactor are equal to those in the effluent stream. When cell growth is limited by a single nutrient, e.g., glucose, glutamine or oxygen, continuous culture is also called *chemostat culture*. For microbial systems, typically the cell density is constant over a wide range of dilution rates and only affected by a concentration change of the limiting nutrient in agreement with a saturation-type growth kinetics as described by the Monod model. Published data on continuous culture of animal cells differ from the expected simple Monod model in many cases suggesting that more complex equations govern the system. A linear decrease of the viable cell density with increasing dilution rates has been observed or a relatively stable cell density at low dilution rates, which decreased toward higher dilution rates as well as the opposite. Other researchers found data in accordance with a simple Monod model.

Microbial systems can be kept at constant cell densities (*turbidostat*) under nonsubstrate limited conditions by controlling the dilution rate. A similar approach was applied to run a cytostat where almost arbitrary cell densities were maintained at similar dilution rates using identical feed medium. Despite the deviating behavior of continuous cultures, steady states regarding cell density and growth rate, product and metabolite concentrations, metabolic and production rates are obtained both in chemostat and cytostat operation after at least three to five times the residence time of the medium in the system.

Typically, dilution rates applied are in the range of growth rates observed during batch culture, i.e., 0.3–1.4 day^{-1} and should not exceed the maximum growth rate to avoid wash-out of cells. Cell densities and product concentrations are in the same order as determined during batch culture. In contrast to batch culture, where transient changes of most parameters are observed, the continuous mode offers the advantage to study the effect of different culture variables under steady state environmental conditions.

Growth and metabolism have been characterized over a wide range of dilution rates for hybridoma cells, CHO cells, and BHK cells. Metabolic shifts toward a more efficient energy metabolism were described at very low nutrient concentrations for hybridoma and BHK cells. Europa et al. observed multiple steady states at similar dilution rates depending on the history of the culture allowing it to adjust to more desirable conditions for antibody expression.

Continuous culture has been used to investigate the effect of different culture conditions on a variety of cellular responses. Examples are the glycosylation pattern of BHK, CHO, and NS0 cells, cell death in hybridoma culture, Bcl-2 overexpression and viability, the effect of dissolved oxygen on cell density, energy metabolism and antibody production, shear sensitivity, cell cycle distribution and productivity of hybridoma cells, the effect of pCO_2. Heidemann determined the K_s-values (Monod constants) of essential amino acids for CHO and hybridoma cell lines in continuous culture. Schmid et al. employed the evolutionary potential of continuous culture to adapt hybridoma cells to increased shear forces.

Although continuous culture has proven to be a powerful research tool it is not very suitable for the commercial production of therapeutic proteins. There is a risk of rather unstable expression after a certain period in culture. The selection of low producer cell clones due to their frequently increased growth rate is a major disadvantage besides the low cell densities and productivities obtained in continuous culture.

Perfusion Culture

Another type of continuous culture is perfusion culture where the cells are retained in or recycled back to the bioreactor while fresh medium is supplied and cell-free supernatant continuously removed at the same rate. Suspension cells may be retained in or recycled back to the stirred tank or airlift bioreactors using filtration, centrifugation or sedimentation techniques. Other methods are immurement in hollow fiber bioreactors, flat membrane reactors and dialysis reactors, entrapment in porous matrices cultured in fixed and fluidized beds or encapsulation in gelling polymers. Methods applied for anchorage dependent cells include microcarrier culture in stirred tanks equipped with filters or settlers, cultivation on porous microspheres or other growth substrates in fluidized and packed beds, culture in ceramic matrices and plate bioreactors.

During perfusion operation the cell density increases steadily according to the chosen medium perfusion rate D until nutrient or oxygen limitation or waste product inhibition occurs and a quasi steady-state with respect to cell, metabolite and product concentration is reached. Typically, cell densities in the range of 5×10^6 to 5×10^7 mL^{-1} have been reported for homogeneous suspension cultures and even up to 5×10^8 mL^{-1} for hollow fiber systems. Thus, the engineering challenge regarding oxygen and medium supply is considerable. Advantages of this technique are the high cell densities and product concentrations obtained leading to increased space-time yields when compared to batch, chemostat and even fed-batch operation. It is difficult to make a general statement on the achievable productivity increase since it is dependent on a series of parameters related to the cell, the medium and the applied system. Typically, a productivity increase in the range of 5–10-fold is found in the literature.

Although perfusion is more complex than batch culture and therefore considered to be more prone to contamination it has become an established technique for large scale commercial production of recombinant proteins. Under certain circumstances, e.g., when growth associated product formation is observed or when product formation is dependent on the cell specific perfusion rate, it is better to control the growth rate and cell density by applying a cell bleed. Another advantage is that the accumulation of dead cells and debris and associated negative effects such as release of proteolytic enzymes or DNA are prevented which is otherwise observed in membrane based cell retention systems.

Growth, metabolism, and product formation during perfusion with a controlled cell bleed has been investigated by a series of researchers. Frequently, a natural cell bleed is obtained in packed and fluidized bed cultures or during spinfilter, settler or centrifuge based perfusion of suspension cells. The productivity of hybridoma perfusion systems was further increased by the application of fortified or nutrient enriched media feeds or by a separately controlled substrate feed to minimize toxic metabolite production.

Selection and Design of Bioreactors

There is no single, superior bioreactor suitable for all applications. The bioreactors or cultivation systems describe offer different advantages and their applicability is limited for various reasons. Therefore, some considerable thought has to be given to the selection of a specific bioreactor type and its design. Literature precedents may provide some guidance, however, they should be reviewed carefully and alternatives should be investigated. It is a time consuming and expensive task to undertake an experimental evaluation to compare the relative performance of competing culture systems for a specific application. A range of technical, biological and regulatory issues govern the decision on whether inherent characteristics or complexities would make the system risky for a commercial purpose. In many cases, especially in a nonindustrial setting the choice of reactor system is mainly influenced by the familiarity of the operator with the technology, the availability of space and services; and the investment and operational costs. For industrial applications, homogeneous culture of suspension cells, cell aggregates or adherent cells on microcarriers in STRs can be considered as the predominant technology due to their scale-up potential, flexibility and environmental control features.

The grouping is somewhat arbitrary since one aspect may belong to different categories and parameters are related. It is recommended to start off with a clear definition of the purpose of the bioreactor and potential future changes to the initial framework of conditions. The next step in the process of bioreactor selection and design is to compile a requirement specification list for the application under consideration taking into account the users specific circumstances. Such a document could simply follow the different issues and the level of detail will be governed by the expertise of the team involved with the task.

Important general issues independent of the reactor type are the availability and quality of service. The need for service can vary with the circumstances. A large company with many bioreactors of the

Table 6.1. Recommendations for the mechanical design of stirred tank cell culture bioreactors

Materials	Stainless steel 316 L (1.4404) or other high alloy grades for improved chloride corrosion resistance
	Borosilicate glass for vessels or sight glasses
	Silicone rubber and EPDM (ethylene propylene diene monomer) for gaskets
	PTFE (polytetrafluoroethylene) for electrode sleeves, support bearings
	Ceramics and graphite for mechanical seals
Finishes	Electropolished surface, $R_a < 0.4$ μm (R_a: arithmetic mean roughness)
	Avoid internal threads
Welding	Automated welding methods such as orbital welding
Aspect ratio	1:1 to 3:1, typically 2:1
Impeller	Axial flow impellers such as marine type, large pitch bladed, segment impeller
	Impeller diameter ≥ 0.5 vessel diameter
Sparger	Ringsparger (pore size ~ 1 mm), microsparger (pore size 5–10 μm) or bubble-free via silicone tubings

same type probably might have a competent internal technical support. The initial availability and quality of steam, pressurized air, process gases, electricity, cooling water and drain limits the choice of systems or at least impact strongly on the design. Therefore, hollow fiber bioreactors are frequently found in standard laboratory facilities since these systems are designed accordingly. However, most bioreactor types are available in designs appropriate to limited media and electricity supplies.

Sterile Operation

The most important requirement for cell culture bioreactors is long-term sterile operation due to the slow growth of animal cells (maximum population doubling times in the order of a day are typical). Appropriate sterile design is very important for continuous processes, which may be conducted for periods of months to up to half a year. Small-scale equipment made of glass is generally autoclaved whereas larger bioreactors made of stainless steel are commonly in-situ steam sterilized. Especially, the design of seals, valves, pumps and transfer lines and the sterilization procedures and sequences have to be considered. Double mechanical seals are used for rotating shafts or alternatively magnetically coupled systems. In the sterile, cell and product containing part of the bioreactor, membrane valves are applied as well as peristaltic pumps, which are also appropriate for cell recycle. For successful sterilization complete drainage of condensate and removal of residual air has to be ensured to avoid cold spots and air pockets. If presterilized units are used, coupling of reservoirs and transfer of medium and inoculum needs special attention. In particular, the degree of complexity of a bioreactor or even bioprocess plant has a large impact on a careful sterile design.

Economic Issues

Even though many people feel that economic issues should not determine the selection of a certain bioreactor system it may be the ultimate criterion for the decision. Here, short-term issues such as the investment cost have to be balanced carefully to rather long-term issues such as running costs, achievable space-time-yields and if the bioreactor shall be used for different purposes. Other long-term considerations are time-to-market for a biopharmaceutical or time-to-patient for a medical application, which are affected by the time needed for validation and the probability of process related difficulties during clinical tests.

Process-Related Issues

For industrial applications sterilization-in-place and cleaning-in-place, the degree of process automation, appropriate process control and monitoring of critical culture parameters as well as compatibility to up- and downstream operations are important issues for the choice of a reactor system. Furthermore, downtime and reliability of the bioreactor and process are other critical factors. These parameters are mainly dependent on the complexity of the operations, the operator skills and training. In a commercial context, scalability of the culture system has to be emphasized in particular due to the frequently enormous production capacities required to secure supply to the patients. Yearly production capacities of ten to hundred of kilograms were estimated for therapeutic monoclonal antibodies depending on the dose. Other examples demanding large production capacities are recombinant therapeutic proteins such as tPA, EPO, blood factor VIII, and vaccines used in broad vaccination programs.

Regulatory Issues

For the production of biopharmaceuticals regulatory guidelines affecting the design and operation of bioreactors have to be considered. Issuing authorities are the Food and Drug Authority in the United

States, the EMEA in Europe, and national regulatory authorities. The international conference of harmonization aims to provide one single guiding document, which is independent of the country. These guidelines cover equipment construction, installation and operational qualification and finally process validation where in a number of consecutive runs the consistency of the process has to be demonstrated via a reproducible quantity and quality of the product. However, they cannot be read like a technical instruction manual, instead an assessment of the specific application has to be made. Concerning equipment construction the 21 Code of Federal Regulations, parts 210 and 211 states that "Equipment shall be constructed so that surfaces that contact components, inprocess materials, or drug products shall not be reactive, additive, or absorptive so as to alter the safety, identity, strengths, quality, or purity of the drug product beyond the official or other established requirements". Relevant technical guidance documents on pharmaceutical engineering are provided by the "*Society for Pharmaceutical Engineering*".

Process safety regulations govern the contamination control of the product with impurities such as residual host cell protein, DNA, microbes and adventitious viruses. In multipurpose facilities, cross-contamination caused from previous production campaigns of other substances have to be considered. A major focus concerning the mechanical design of a bioreactor or plant in this regard is on sterility and cleaning, i.e., removal of contaminants as proven during validation. More recently, the use of disposable materials and bioreactors to circumvent contamination problems and minimize validation costs has been discussed.

Other important aspects related to regulatory requirements are product degradation, truncation, and modification, which may be affected by the choice of the culture system and the process parameters. An example is the potentially increased product degradation in high cell density systems with product accumulation due to the increased concentration of proteases and the eventual release of intracellular proteases due to cell death and subsequent lysis. Proteolytic degradation of complex proteins has been observed in cell cultures of BHK cells known for their high proteolytic activity. Additionally, the genetic stability of a production cell line will impact on the choice of the process mode and culture conditions and therefore indirectly affects the selection of a bioreactor type. This is especially important for continuous operation when growing cultures in perfusion and chemostat mode. One clear advantage in this respect are growth arrested continuous cultures, which can be obtained in high cell density systems without cell bleed or via changing the culture temperature, adding chemicals, or using genetically engineered cell lines.

Biosafety Issues

Biosafety assessment of the cell line and process will have impact on the containment design of the culture to control the risk to health or environment posed by the genetically modified organism. Biosafety regulations or guidelines are issued by the National Institute of Health in the United States, the World Health Organisation WHO, the European Commission, and different national authorities. These regulations are relevant for the design of seals, valves, air vents, sampling ports, addition ports, and other parts of a bioreactor where a potential risk for release of hazardous organisms can be assumed. No particular design solutions are prescribed since the focus is on performance of the containment design in order to allow for technical improvement. Detailed technical information valid for the EC is found in the CEN standards which are a technical complement to the legislation. Some containment design considerations for bioreactors are described in.

Cell Line and Product-Related Issues

Probably better covered in the scientific literature are design and selection considerations related to the cell type and product class. A basic demand is appropriate temperature control. This needs special attention in the case of cells passing through external loops where control of the environmental conditions can be poor dependent on the design and the residence time of the fluid in the loop. Commonly, mammalian cells are grown at 37°C and insect cells at 27°C. However, the beneficial effect of a decreased culture temperature has been shown for various mammalian cells. Other temperature control profiles have been suggested to control growth and productivity. It is known that membrane fluidity of mammalian cells is temperature dependent and therefore a temperature effect on shear sensitivity cannot be excluded. All this emphasizes the importance of a reliable and precise control of culture temperature.

Parameters affecting the shear sensitivity of animal cells include the concentration of serum, proteins, and surfactants. Although the shear tolerance is cell line dependent and encapsulation in gel-forming polymers or entrapment in macroporous carriers may protect the cells or reduce the accessibility it clearly affects the design and operation conditions of stirred tank, airlift, fluidized, and packed bed bioreactors.

The generated shear rates depend on the design and operation parameters of impellers and pumps and in particular the sparging conditions as outlined in different sections of this chapter. Bubble-free aeration is gentler, however scale-up is difficult regardless the system used. Due to the intense research and development related to shear force generation, operation conditions and bioreactor configurations have been identified that are suitable for large-scale gentle cell cultivation in stirred tank, airlift and fluidized bed bioreactors. In parallel, other systems have been developed where the cells are not exposed to mechanical forces, e.g., hollow fiber bioreactors, the static maintenance bioreactor, and the membrane bioreactor. The quantification of the shear sensitivity of different cell lines under various process conditions in combination with practical experience showed clearly that animal cells are not at all that sensitive to mechanical forces as originally assumed.

The minimization of mechanical forces is not the only design criterion for oxygenation systems. CO_2 removal and sufficient oxygen transfer are other important criteria for the selection of a bioreactor system. Controlling the oxygen partial pressure in cell cultures is of utmost importance for cell growth, productivity and metabolism as shown by many authors. In general, oxygen supply is a major engineering challenge in heterogeneous system, e.g., fluidized bed, packed bed and solid bed bioreactors, and high cell density systems, e.g., hollow fiber systems, perfusion and fed-batch bioreactors. The oxygen transfer characteristics and limitations of the different systems are reviewed in the individual sections describing the reactor type. Different technical solutions for improved oxygen supply are also discussed. Related to oxygen supply is also the control of an appropriate culture pH. Generally, mammalian cell culture media are buffered using the bicarbonate buffer system and the culture pH is affected by CO_2 and lactate formation. If base addition is used to compensate for acid formation a concurrent increasing osmolality is observed. It has been shown that the culture pH, CO_2, and an increased osmolality may affect productivity and glycosylation. Therefore, for sensitive cell lines a culture system should be chosen allowing for pH monitoring and control via addition of base or ventilation of the culture system avoiding major osmolality changes. Beside temperature, pH, and pO_2 control, the supply of nutrients and removal of waste products affect cell cultures. Significantly improved product yields are typically obtained from continuous perfusion and fed-batch processes in comparison to simple batch mode.

Product feedback inhibition needs consideration in high cell density culture since it will prohibit the application of high cell density systems where the product is accumulated in the cell containing compartment (e.g., hollow fiber systems, microencapsulation, membrane filtration in combination with fouling or fed-batch systems). The product localization, i.e., secreted into the medium, membrane bound or intracellular, and the nature of the product, i.e., produced by the cells or the cell itself, is affecting bioreactor design and selection. For the production of secretory products continuous perfusion operation of fluidized and packed bed bioreactors provide an interesting alternative to conventional stirred tank technology. If the cells need to be harvested for further processing easy accessibility and quick, safe and reliable harvest procedures are advantageous. The product formation kinetics dictates the optimum operation mode and therefore indirectly the selection of a bioreactor type, since not all reactors are suitable for a given mode.

The cell type, i.e., anchorage dependent or suspension, has significant impact on the design and selection of an appropriate bioreactor. Although a series of bioreactors can be modified to be suitable for the cultivation of both cell types (e.g., stirred tank, airlift, fluidized bed with or without different types of solid or macroporous carriers) there are applications that require a certain bioreactor type since significant improvements in productivity are obtained, or the cell line cannot be grown or do not show the required functionality in a "*multipurpose*" culture system. A prominent example is artificial organs where a certain three dimensional architecture is required for an organ-like behavior. Especially in the case of a single purpose bioreactor one would decide for the most appropriate type and design for the specific application, i.e., yielding optimum productivity and product quality. However, true multipurpose design is required when different cell lines are cultured for the production of various intracellular, membrane bound or secretory products, e.g., in drug discovery using recombinant proteins for structure determination or as tools for high throughput screening of novel drugs. Other applications that benefit from multipurpose design are process development and pilot production for different classes of biopharmaceuticals.

Although a series of growth and productivity promoting serum-free or even protein-free media have been developed during the last 15 years not all processes are conducted at low protein concentrations. This is due to the requirement of a particular cell line for a certain protein concentration and/or availability of growth factor(s) or simply the reluctance to adapt the cell line to low protein serum-free conditions. As the viability decreases cellular proteins and DNA are released in the culture supernatant due to cell lysis. The increasing concentration of proteins and DNA are causing membrane fouling and reduced performance of membrane based cell retention systems. Therefore, the use of centrifuges or settlers is frequently preferred. Another important aspect using low protein serum-free media is the increased chemical reactivity of such media especially in the presence of high chloride concentrations degrading certain stainless steel qualities. In conclusion, the choice of a bioreactor system and its design is very individual and needs careful consideration of the specific demands of the application.

Concluding Remark

A wide range of reactor types has been suggested in the literature for animal cell culture including stirred tanks, airlift reactors, fluidized bed reactors, packed bed and solid bed reactors, hollow fiber and membrane reactors. The number of different types is even increased for small-scale simple systems such as spinner, roller, and shake flasks. Typical operation modes are batch, fed-batch, and perfusion although the suitability is strongly dependent on the application, the cell line, and the product. For certain bioreactor systems a broad applicability for different cell lines and purposes has been demonstrated. However, there is no single cell culture system that is universally suitable for all applications in cell culture technology. For commercial production of recombinant proteins, monoclonal antibodies, or vaccines, homogenous culture of suspension cells, cell aggregates, or adherent cells immobilized on microcarriers in SRTs operated in batch, fed-batch, or perfusion mode can be considered as the predominant technology due to its maturity, scale-up potential, flexibility, and environmental control. The specialist reactor designs are experiencing a renaissance at present within the field of artificial organs and tissue culture for replacement therapy. At the same time, conditions for using the different suggested reactor designs for the production of viral vectors for genetic vaccination or gene therapy are being investigated. Apart from these novel applications bioreactor development has reached a high degree of maturation and the focus of cell culture technology has moved to biological issues and novel applications of cell and tissue culture.

7

OPTIMIZATION OF CELLULAR POPULATION

This chapter discusses the benefits and the techniques of high cell density perfusion bioreactors for culturing mammalian cells. Perfusion refers to the continuous in flow of nutrient medium coupled with the out flow or harvesting of spent medium, while the cells are fully or partially retained within the bioreactor In contrast to the chemostat bioreactor, in which the out flow stream contains cells at the same concentration as in the bioreactor, the out flow stream of a perfusion bioreactor is either completely or substantially depleted of cells. The concept of perfusion was applied as early as 1912 to keep small pieces of tissue alive for extended periods of observation under a microscope. A historical account on the development of perfusion bioreactors for mammalian cell cultivation is given in an earlier review. While many of the perfusion techniques discussed earlier pertain to attachment cultures and are still applicable, this chapter focuses more on perfusion techniques for suspension bioreactor cultures, as these homogeneous cultures have become increasingly popular in the ensuing decade.

CASE FOR PERFUSION BIOREACTOR CULTURES

In addition to batch and fed-batch bioreactor cultures, which are traditionally preferred for the production of the therapeutic proteins in recombinant bacterial cells, chemostat and high cell density perfusion bioreactor cultures are becoming increasingly acceptable for culturing of mammalian cells due to a several key advantages. First, continuous addition of fresh nutrient medium and removal of spent medium provide an easy method of removing the metabolic by-products, ammonia and lactate, which are toxic to cells if allowed to accumulate as in batch or fed-batch cultures. Second, the facile secretion of recombinant proteins from mammalian cells (compared to intracellular hoarding of the heterologous proteins by bacteria) eliminates the need for cell harvesting at the end of a batch or fed-batch process, and cell disruption to obtain the secreted protein product. Third, due to the long doubling times involved in the growth of mammalian cells, it is more efficient to retain the viable and productive cells inside the bioreactor rather than growing the cells repeatedly in batch cultures. Fourth, even partial retention of viable and productive mammalian cells inside the bioreactor results in higher cell density compared to the low cell density batch or chemostat cultures. As the protein synthesis rate in the bioreactor is directly proportional to the number of viable and productive cells, the higher cell density in perfusion bioreactors results directly in a higher protein production rate. Next, at the low cell *growth* rates prevalent in the high cell density perfusion cultures, many hybridoma cell lines exhibit higher specific (per cell) production rates, resulting in much higher protein production rates. Finally, the shorter residence time of the product proteins inside the perfusion bioreactor is advantageous in minimizing their exposure to proteases, sialidases and other degradative enzymes that are released into the culture medium from the accumulating dead cells in the fed-batch bioreactors.

As a note of caution, monoclonal antibody production in long-term continuous bioreactor cultures of hybridoma cells has sometimes been found to be unstable. However, this gradual instability of antibody production in the slower growing hybridoma cultures is in marked contrast to the rapid instabilities observed in heterologous protein expression from recombinant bacterial cultures. With the slower dynamics of protein production loss in mammalian cells, perfusion bioreactor cultures may still be sufficiently stable over several weeks or months of continuous operation and result in much higher protein production over the repeated batch cultures in the similar sized bioreactors. Thus, perfusion bioreactor cultures represent a superior bioreactor operation for large-scale production of therapeutic proteins from mammalian cells, if the instability of the specific protein synthesis rate is not a serious issue.

Industrial Application of Perfusion Bioreactors

There are a number of products in the market derived from cell culture technology. A greater number of products are in development and it is very likely that some of these will be approved for commercial production. The demand for cell culture–derived products is high in most cases and the manufacturing cost is significant. Some of the antibodies, for instance, are administered at high doses and a production capacity of 500–1000 kg/year is required to meet the demand. There is no question that the cell culture processes must be optimized with respect to overall yield and productivity to meet these demands. As discussed, one of the key advantages of perfusion bioreactors is the higher protein production rate. Due to high cell densities achievable in a perfusion bioreactor, the volumetric production rate can be 4–10 times higher as compared to a fed-batch bioreactor. This higher production rate allows the cell culture process to utilize compact bioreactors, which are easier to operate and control. Product quality is another key advantage for perfusion bioreactors especially when the product is labile and prone to degradation in the bioreactor. Even though the batch or fed-batch operation is widely accepted by many companies, these key advantages resulted in perfusion processes for many products.

The first industrial scale perfusion process was developed by Centocor in the late 1980s for the production of Centoxin, an antibody for the treatment of sepsis. In 1991, the drug was approved for use in Europe. Centoxin failed the clinical trials in the United States and was discontinued. Centocor later developed the process and obtained approval for ReoPro in 1994 using perfusion technology. The same year Genzyme obtained approval for Cerezyme produced in a perfusion process. Other perfusion culture products like Gonal-F from Serano, Remicade from Centocor, ReFacto from Wyeth, Kogenate-FS from Bayer, and Zigris from Eli-Lily reached the market in quick succession. The products from the perfusion process have a sizable market. The sales of perfusion-based drugs are significant; they are multimillion dollar drugs, some of them blockbusters. Remicade is a monoclonal antibody produced at a rate of several hundred kilograms a year with sales reaching $1.5 billion/year.

Cell Retention Systems

The key feature in operating a perfusion bioreactor is substantial or full retention of cells inside the bioreactor. The methods for retaining cells inside the bioreactor are primarily determined by whether the cells are growing attached to surfaces or growing in either *single cell suspension* or *cell aggregates*. While most mammalian cells historically were grown attached to a surface or a matrix, the difficulties in scaling up these surfaces for the production of large amounts of therapeutic proteins have encouraged efforts to adapt many industrial host cell lines to grow as suspended cells. With the increasing success of adaptation of these cells to grow in suspension, the previously developed methods of perfusion cultures for attached cells are now falling out of favor. Nevertheless, for any cells that are difficult to adapt for growth in suspension, the traditional techniques for perfusion cultures are summarized below.

Cell Retention Systems for Heterogeneous Cultures

The different heterogeneous systems summarized below were developed for retaining mammalian cells inside the bioreactor through methods of cell attachment to surfaces and entrapment of cells in different devices. The heterogeneity of these systems arises from the segregation of cells into defined compartments within the bioreactors, i.e., the cells are not uniformly distributed throughout the culture volume in the bioreactor.

Fixed (immobilized) bed

Various matrices (e.g., glass beads, sponges, fibers, etc.) are typically packed in smaller vessels and inoculated with attachment-prone cells for the production of viruses and proteins. Glass beads are the most popular matrices and they range in size from 3 to 5 mm. These perfusion bioreactor vessels have been scaled up to 100 L, but are usually much smaller. While lytic viral production is typically accomplished in batch processes, Brown et al. have used glass beads as the attachment matrix for attachment-prone mammalian cells in perfusion bioreactors of volume up to 60 L for periods ranging from 4 months to over 1 year. Scale-up of bioreactors is typically limited by the weight-carrying capacity of glass beads, the uneven inoculation of cells onto the matrix in a deep bed, and poor oxygen transfer.

Ceramic matrix immobilization (opticell)

The *Opticell* system using a ceramic matrix with either a nonporous or porous surface has been used for the perfusion culture of both suspension and attached cells. The porous surfaces can entrap suspension cells and provide a larger surface area for the attached cells, while nonporous smooth surfaces are suitable for harvesting cells. The system uses a computer to control the common bioreactor parameters, such as pH, oxygen, and perfusion rates. Inoculation of the complete matrix requires a four-step process to ensure that all the channels of the cylindrical ceramic matrix are loaded with cells. Scale-up of these ceramic cartridges is limited by the gradients in nutrient concentration along the length of the cylinder.

Hollow fiber reactors

Hollow fiber cartridge systems have been quite successful as small-scale perfusion bioreactors for the production of monoclonal antibodies from hybridoma cultures. Ultrafiltration capillary fibers enhance the antibody concentrations along with the high density of hybridoma cells in the extracapillary space, while nutrients and waste products are easily exchanged with lumen. Scale-up of these bioreactors has been hampered by the pressure drop as well as spatial variations in nutrient and waste product concentrations along the length of the reactor. Several other flow patterns have also been explored to overcome some of these problems to different degrees of partial success.

Microencapsulation of mammalian cells

Encapsulation of living tissues inside polymeric semipermeable microspheres has been developed for potential biomedical applications and easily adapted for mammalian cells as well. The porosity of the polymer layer can be controlled to allow the permeation of smaller nutrient and waste metabolite molecules while retaining the larger protein product inside the microcapsules. The encapsulation technology is based on polyionic binding of poly-L lysine, poly-L orinithine, poly-L glutamate or other basic polymers with gelled calcium alginate bead containing the mammalian cells, followed by liquefaction of the capsule interior. Another polymeric material used in microencapsulation of mammalian cells is chitosan biopolymer, which coacervates onto the gelled alginate beads. Size of the gelled beads can be controlled in the range of 100–500 μm by using different syringe sizes as well as application of an electrostatic field. A claimed advantage of concentrating the protein products inside the microcapsules is countered by the possible degradation of the protein by the proteases excreted by the dead cells accumulating inside the microcapsules. Recovery of the accumulated product from the capsule interior presents another complex operation.

Macroporous matrix, fluidized systems

Microcarrier culture systems can be easily used in perfusion bioreactors, as it is very easy to separate the microcarrier beads from the harvest stream. The microcarrier beads introduced by Van Wezel et al. provide a surface matrix for the attachment and growth of attachment-dependent cells. These beads are suspended in a homogeneous nutrient environment through gentle agitation. Porous microcarrier beads increase the surface available for cell attachment and growth. Two porous beads that have been used in perfusion cultures of attachment-dependent mammalian cells are the *Verax collagen beads* and the *gelatin beads*. With the ease of adapting many industrial host mammalian cells to growth in suspension cultures, these processes are no longer in vogue in comparison with the methods of cell retention in homogeneous systems discussed below.

Cell Retention Systems for Homogeneous Cultures

In homogeneous systems, mammalian cells are grown in suspension cultures and the cells are distributed uniformly throughout the bioreactor culture volume. A major advantage of using a suspension culture is its scalability and the ability to monitor and control results due to the relatively uniform environment. While suspension cultures are naturally used for attachment of independent cells like hybridoma and myeloma cells, the increasing success of the suspension adaptation protocols for other mammalian cells used in the industry (e.g., CHO and BHK) makes it possible to use the scalable suspension bioreactors for these cells as well. Following the previously established paradigms of recombinant microbial cultures and mammalian viral production systems, batch and fed-batch processes are also commonly used for the large-scale production of therapeutic proteins from mammalian cell cultures. However, in light of the given earlier, it is gradually becoming common-place to consider the continuous perfusion cultures for the production of therapeutic proteins if the cells are sufficiently stable.

The key difference between a chemostat and a perfusion bioreactor is the substantial or full retention of the freely suspended cells inside perfusion bioreactors. Higher cell densities in the perfusion bioreactor are achieved through partial or full retention of suspended cells from the harvest or outlet stream back to the bioreactor. There are several methods for retaining cells inside bioreactors or recycling the cells back to the bioreactor. These methods are discussed qualitatively in this chapter, with an emphasis on optimization and scale-up potential of these retention systems for industrial scale bioreactors.

Spin filters

The earliest cell retention device used for mammalian cell cultures is the *spin filter*, developed by Himmelfarb et al. in 1969 and successfully scaled up to larger (40 L) bioreactors by Tolbert et al. A spin filter is a cylindrical membrane or stainless steel mesh of defined pore size, usually attached to the impeller shaft of the bioreactor. The

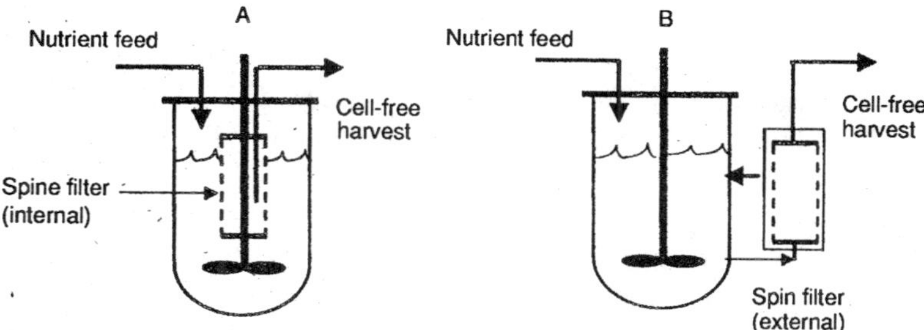

Fig. 7.1. Spin filter for retention of mammalian cells for a high density perfusion bioreactor.

cell-free culture medium passes through the filter and is pumped out of the bioreactor, while the cells are retained inside the bioreactor. With successful cell retention by the spin filter, higher cell densities (typically an order of magnitude higher) are achieved easily in early spin-filter perfusion bioreactors, resulting in a similar increase in bioreactor volumetric productivity. Contrary to a chemostat culture, the perfusion culture with a spin filter is an unsteady state process, with the total and viable cell concentrations not reaching a true steady state unless a cell bleed stream is removed from the bioreactor.

The perfusion culture duration is significantly longer than that of a fed-batch culture process, however clogging of the spin filter leads to the eventual termination of the perfusion bioreactor. The operating life of the spin filter perfusion bioreactor is mainly a function of the effective pore size of the filter and is also affected slightly by the filter materials and rotational speed. With a 5 μm pore size, the spin filter for a hybridoma perfusion culture, clogged within 7 days. With a larger 15 μm pore stainless steel mesh spin-filter, Deo et al. operated a 500 L continuous perfusion bioreactor for 30 days with the mesh clogging increasing with time. Yabannavar et al. increased the pore size further to 25 μm–significantly larger than the mean size of a single mammalian cell; they and found a significant reduction in cell retention efficiency. With the larger 50 μm pore size filter, there was no detectable retention of single cells.

Avgerinos et al. examined the effect of filter materials on fouling in perfusion cultures of a CHO cell line grown initially on microcarrier beads of diameters ranging from 115 to 195 μm and gradually forming aggregates between 200 and 600 μm in diameter. Stainless steel mesh filters ranging in pore size from 44 to 105 μm became plugged within 11–21 days, while fouling was not a significant problem for up to 54 days with a hydrophobic polymer (ethylene-tetrafluoroethylene) membrane. Esclade et al. suggested that the high surface charge density of metals compared to a polyamide membrane might be responsible for the observed differences in fouling of these materials over a week-long perfusion run. However, durability, autoclavability, and reusability of the stainless steel mesh filters make them more attractive than the polymeric membranes.

Deo et al. have further identified rotational velocity as another key parameter affecting the possible maximum perfusion rate before filter clogging. The rate of filter fouling increases with the perfusion rate and cell concentration both. Even with these limitations, spin filter technology has been scaled up successfully to a perfusion bioreactor volume of 500 L in the reported literature. Because replacement of internal spin filters is impractical, spin-filters should be designed with the appropriate choice of pore size and materials, and operated at an optimal rotational speed to prolong the high cell density culture operation before filter clogging eventually forces the termination of perfusion bioreactor.

External filtration

Considering the difficulties in removing the clogged spin filter from the bioreactor without terminating the perfusion culture, it may be advantageous to move the cell retention function to an external filtration module. Cross-flow micro filtration is the preferred method for continuously removing the cell free harvest on the permeate side while returning the concentrated cell suspension to the bioreactor. Both flat-plate and hollow-fiber cartridges have been used traditionally as the filtration devices for this external filtration step. Here again, clogging of the filter with time becomes a serious issue. However, it is possible to replace the external filter cartridge when clogging reduces permeate flow significantly, typically between 5 and 7 days.

Majority of reported studies used microporous membranes with 0.2–0.65 μm pore size, with a few studies also using 5 and 10 μm pore membranes. While small pores are expected to clog more quickly than larger pore filters, the decreased cell retention efficiency (<70%) of the 10 μm pore membrane places an upper limit on the pore size that can be used. Permeate flux is another key parameter affecting rate of fouling, with faster fluxes drastically reducing the time of clog-free operation. Increasing the flow rate of cell suspension through the filtration device reduces the membrane fouling rate; however, higher flow rates result in higher shear rates at the membrane surface. High shear rates have been shown to cause significant cell damage and a drop in cell viability.

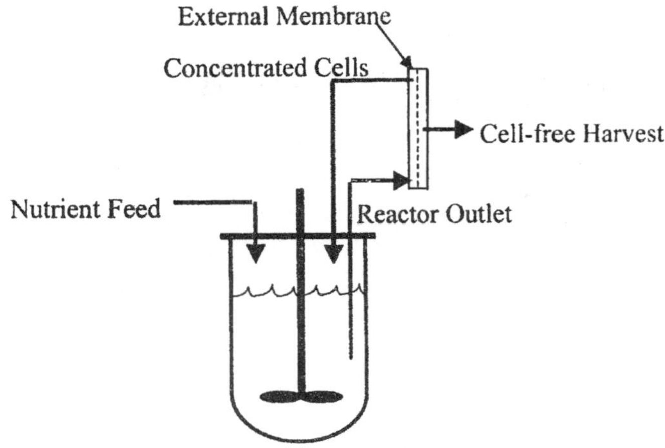

Fig. 7.2. External filtration for retention of mammalian cells.

Some recent efforts are underway to reduce fouling through Taylor vortices in an external vortex flow filter .These efforts have extended the culture duration to 45 days with 0.8 μm pore vortex flow filter 57 and 60 days with a 10 μm steel mesh vortex flow filter. However, all the published results using external filter cartridges were conducted in the smaller laboratory scale perfusion bioreactors (less than 4 L reactor volume), highlighting the difficulties encountered in scaling up the two-dimensional filtration surface area necessary for retaining cells and recycling them back to the bioreactor. Persistent problems with filter clogging have been avoided successfully by the three cell retention methods discussed next.

Alternating tangential flow (ATF) system

Filter fouling and clogging problems with the filtration-based cell retention system may be reduced by ATF through the filter. In the ATF system developed by Refine Technology a 0.2 μm hollow fiber cartridge is connected to the bioreactor by a single port. Cell suspension is alternatively pumped into the filtration module and back to the bioreactor by a fast diaphragm pump. The pump is partitioned into two chambers by a flexible, medical-grade silicone diaphragm. The controller cycles filtered air to and from one of the pump chambers and a positive or negative pressure gradient is produced relative to the bioreactor. The unit is attached to the bioreactor using a single port and a cell free harvest is pulled out from the permeate of the hollow fiber cartridge. The system can be scaled up based on the surface area of the hollow fiber cartridge. Based on the filter sizes available in the market it is possible to run the unit at more than 1000 L/day perfusion rates. The ATF system is designed for suspension cells but a slight modification can make it work with attachment-dependent cells. In anchorage dependent, microcarrier-based cultures, the modular nature of the ATF system allows replacement of the hollow fiber with a *screen module* (SM). This converts the ATF system to a highly effective perfusion device for microcarrier-based cultures. The screen module retains the microcarriers in the system as the microcarrier free medium is removed continuously for perfusion.

Cell sedimentation

Vertical sedimentation

Gravitational sedimentation of cells in quiescent liquid volume has been used to clarify the culture broth in a separate cell settling chamber, to remove the clarified liquid as the harvest stream, and to return the settled cells to the bioreactor. The settling chambers uses gravity to settle and separate the cells. While settling of the cells can be carried out in any geometry, a conical shape offers some advantages for *vertical sedimentation*. Due to increased diameter in the cone, the linear liquid velocity becomes lower than the settling velocity of the cells and the cells settle. Kitano et al. demonstrated the utility of a vertical sedimentation device placed above the bioreactor, with a single pipe transporting cells from bioreactor to the settler and the settled cells from the settler to the bioreactor. Such devices have also been used by other researchers to achieve high cell densities in small-scale perfusion bioreactors. Hulscher et al. improved the piping of the settler, by separating the inlet to the settler and the settled cell recycle to the bioreactor, and demonstrated a complete recycle of viable cells and selective removal of only dead cells in the settler harvest stream. However, these vertical sedimentation devices have a narrow range of upward flow rates, which can be used to clarify the cells from the harvest stream. Hence, the scale-up of the vertical sedimentation devices to larger scale bioreactors has not been successfully demonstrated. The size of these devices at high perfusion rates needs to be huge to effectively separate mammalian cells in suspension with low settling velocities. The cone settlers can be used at a large scale for microcarrier, based bioreactor systems however. In this case, settling is used to keep the microcarriers in the bioreactor during perfusion.

Inclined sedimentation

Inclined sedimentation overcomes the limitations of vertical sedimentation in terms of efficiency and scalability. In these systems, cell suspension is forced to flow up between a pair of inclined plates. The cells settle through a short vertical distance between the plates and move down the surface of the plates to be returned back to the bioreactor. The settling efficiency in inclined sedimentation systems is enhanced due to the Boycott effect. In addition, the cell retention capacity of inclined sedimentation devices is easily increased by scaling up the settling area by increasing the width and length, and by stacking a number of plates over each other. The systems can be made very compact by decreasing the distance between the plates. Although inclined sedimentation is an established method for enhanced particle separation in other industries, its application to cell culture technology is relatively new. Early results on its application to perfusion bioreactors were presented in the late 1980s and these systems had been fully developed for commercial use in the next decade.

Batt et al. first demonstrated the use of an inclined sedimentation device as a *selective* cell retention device to achieve high cell densities in a perfusion bioreactor. Larger live hybridoma cells settling on the bottom plate of the settler roll down this plane and are returned to the bioreactor, while the smaller dead cells and cell debris do not completely settle and are removed in the settler over flow stream. Because of the continuous removal of dead cells and cell debris, this perfusion bioreactor can be operated at a steady state for longer periods without any buildup of dead cells or cell debris inside the bioreactor. Due to the selective recycle of live cells from the settler to the bioreactor, high cell densities (over 10 million cells/mL) are easily achieved in the perfusion bioreactor. With the reduced vertical sedimentation path in the inclined settlers, much higher perfusion flow rates can be used in these devices, overcoming the limitations of the vertical sedimentation devices. The inclined settlers have been scaled up to 40 L and to 500 L perfusion bioreactors through stacking up of multiple inclined settlers, which are known as the lamellar settler. At these production scales, the inclined sedimentation devices have exhibited greater than 95% cell retention efficiency for over 20 days of high cell density perfusion culture.

Searles et al. have adapted the inclined settler, for the perfusion culture of attachment-prone Chinese hamster ovary cells. By increasing the settler inlet flow and liquid recycle to the bioreactor independent of the harvest or perfusion rate, the cells that settled in the device are continuously swept back into the bioreactor. Searles et al. have shown that more than 90% of the live cells entering the settler are returned to the bioreactor at the same residence time as the liquid residence time in the settler. The small fraction of live cells (between 5 to 10%) that enter and form a sediment layer in the settler are easily dislodged and resuspended into the liquid flowing down to the bioreactor by bubbling sterile headspace gas into the settler as first demonstrated in industrial scale perfusion bioreactors. Searles et al. have also investigated other modifications such as cooling the settler to 4°C to slow down the metabolism of live cells in the settler and release any attached cells, as well as vibrating the settler to dislodge any cells attached on the settler. These strategies have been successfully implemented in industrial scale continuous perfusion bioreactors.

Tyo et al. described a laboratory scale perfusion system based on a conical lamellar settler. This system was incorporated into a spinner flask to grow cells to high cell densities. Knaack et al. extended this concept to design a conical bioreactor with internal settling zone. The bioreactor was shaped as a cone and conical plates were placed on the top to retain the cells in the bioreactor. The design of a lamella settler can be simplified by the use of plates rather than cones. Several investigators presented successful operation of inclined plate settlers. These plate settlers are compact in size can be operated at high perfusion rates. In these systems cell suspension flows up between the plates in laminar flow conditions. While they are in transit, the cells settle on the plates and return back to the bioreactor. Typically, the settlers are angled at 30 from vertical and the plates are separated by 0,5–2 cm distance. The dimensions of the plates are typically 5–15 cm in width and 50–100 cm in length. The number of plates in a settler can vary between 5 and 20 plates. The efficiency of a settler for cell separation can be enhanced by cooling the cells to room temperature or lower, and by applying an intermittent vibration to the plates.

Thus, inclined settlers have been scaled up to production scale perfusion bioreactors to continuously remove dead cells and cell debris and to selectively recycle the viable and productive cells back to bioreactors. While this retention device has no moving parts to damage the cells and is easy scalable, one drawback expressed by the industrial practitioners is that the deceptively simple device is bound by the in flexible performance characteristics governed by the different sedimentation rates of larger live and smaller dead cells.

Centrifugation

Another approach to circumventing the problem of filter clogging is the use of continuous cell centrifuges with rotating mechanical seals.

While this cell retention method has been shown to retain cells at 100% efficiency in 12- and 15-day perfusion cultures, the effect of repeated centrifugation on cell viability remains a concern. The potential failure of the rotating seals over a long perfusion run led to the development of the Centritech laboratory centrifuge with an ingenious tubing connection to enable continuous centrifuging. Johnson et al. have investigated this centrifuge with a small-scale perfusion culture of hybridoma cells, and raised concerns of shear stress and nutrient deprivation in the pelleted cells. Intermittent operation of the centrifuge was suggested as a possible method to overcome these concerns. In addition, serious issues remain concerning the durability of the centrifuge components, such as the air barrier, the centrifuge insert, and the supernatant tubing. Thus, while trying to eliminate the problems of clogging in the simpler filtration devices, the centrifuges introduce the additional problems of more complex mechanical devices and their effects on cell viability. The use of centrifugation was demonstrated for large-scale production of biologicals from mammalian cell culture. The perfusion rates were reported to reach 3600 L /day in these systems and the cell retention efficiency was more than 90%.

Centrifugation cell separation also resulted in interesting bioreactor designs for the culture of mammalian cells. A centrifugal bioreactor is essentially a continuous centrifugation device where the cells are retained in the bioreactor by centrifugation and perfused continually by the incoming medium stream. This system is very compact and has potential for large-scale culture. However, it needs to be scaled-up and issues related to aeration and mixing must be addressed.

Ultrasonic separation

Ultrasonic separation is based on acoustic aggregation of suspended cells in the standing wave field due to differences in density and compressibility between the cells and the medium, and the enhanced cell sedimentation upon the removal of the field. The ultrasonic resonator, is not susceptible to fouling or mechanical failure, as it does not have a physical barrier or moving parts. Trampler et al. have utilized this cell retention device on a 1 L bioreactor in a month-long perfusion culture of hybridoma cells with no detectable impact on cell viability. Cell retention at higher flow rates from a 5 L bioreactor requires a higher power input than the ultrasonic wave generator. More recently, this ultrasonic separation device has been scaled up to retain cells from the harvest stream of a 100 L perfusion bioreactor by housing four of the ultrasonic units operating at even higher power in cooling water. The higher power requirement (90 W) and multiplexing design of the larger device highlight potential problems in further scaling up this cell retention device due to the nonuniformity of the force generated in larger standing wave fields and the removal of heat generated within this device.

Hydrocyclones

Hydrocyclones are used to separate solids from suspension using centrifugal forces generated by a flow field. The suspension enters to the hydrocyclone tangentially to the wall. The solids are concentrated and removed in the under flow and the clarified liquid is removed from over flow. The system is very attractive to perfusion bioreactors because of its small volume and its high separation efficiency. However, operation of hydrocyclone requires high flow rates and a relatively high pressure drop. The use of a hydrocyclone for cell separation was demonstrated first for yeast fermentation). A recent work by Jockwer et al. reported successful use of a hydrocyclone for CHO cell perfusion culture. A small unit with 3 cm diameter could be operated at 500 L/day perfusion rates with cell separation efficiency of more than 70%. The damage to the cells at high flow rates was minimal. The drop in viability was only 5%. The system could be operated at higher separation efficiencies (>90%); however, the drop in viability was higher (11%).

Scale-Up Potential of Cell Retention Systems for Industrial Application

It is clear that there are many options for cell retention and, depending on the application, one system may be a better choice than others. The simplicity in operation, robustness, scalability, and process economics are the factors that help determine the applicability of the system for use in industrial settings. Although a lot of effort has been devoted to their development, cell retention systems for heterogeneous cultures such as fixed beds, hollow fiber systems, micro encapsulation, and immobilized cell systems have proved to be unsuitable for large-scale production. The main reasons for the absence of any encapsulation or immobilization methods in large-scale production processes are the difficulties in scale-up, intensive labor, and the cost involved in cell immobilization. Similarly, the complexity of the operation limits the acceptance of cell attachment systems such as fiber beds, ceramic cartridges, and microporous carriers for commercial production. Most of the bioreactor systems such as OptiCell and the Verax fluidized bed reactor have been abandoned. Only a limited number of technologies such as hollow fibers find application for small quantity production of antibodies for diagnostic purposes.

The cell retention systems for homogeneous systems have a better acceptance record for industrial applications. These systems use internal or external cell retention devices to separate the cells and they are easier to operate and easier to scale-up. The capacity of the cell retention system is measured in terms of cell density, the degree of cell retention, the duration for the operation, and perfusion flow rate. In order to discuss the limits in their performance, perfusion systems can be categorized into two categories: (a) *filtration-based systems* and (b) *open perfusion systems*.

Filtration based systems include spin filters, external filtration, and ATF. Since they are filtration based, the degree of cell retention can be very high and if one excludes the spin filter system, the retention is 100% and does not change with the flow rate. Under limiting conditions the number of cells achievable in the bioreactor will be dependent on the medium composition and the medium exchange rate. Medium exchange rate and medium depth (nutrient concentration of medium) both result in an increase in cell density. The limitation of filtration based cell retention system exhibits itself in run length. These systems eventually clog and either the run must be terminated (for an internal cell separation device) or the device needs to be replaced (for an external cell separation device). The time to clog can vary between 10 and 100 days. A number of factors such as cell density, viability, perfusion rate, filter material, and pore size are involved in determining the time to clog.

Open perfusion systems include gravitational setters, centrifuges, ultrasonic separation devices, and hydrocyclones. As in the case of filtration based systems, the maximum cell density attainable in these systems increases with the perfusion rate and with the medium depth. These open systems, by definition, do not clog so they can be operated indefinitely. The operating limit for these systems is exhibited in the degree of cell retention. Unlike filtration based systems, the degree of cell retention is reduced at high perfusion rates. This reduction is dependent on such factors as cell diameter, aggregation, and other process parameters. The open perfusion systems require a high degree of cell retention, typically more than 90%. The loss of cells in open perfusion systems should be less than the growth rate of the cells for a stable operation. The degree of tolerable cell loss or the minimum degree of cell retention limits the maximum perfusion rate. This maximum perfusion rate then limits the maximum number of cells attainable in the bioreactor.

Cell entrapment systems can achieve very high cell densities (>50 million cells/mL) and they can be operated for a long time. The 3-D microcarrier system developed by Verax could operate for more than 6 months at perfusion rates close to 600 L/day. However, this system, along with other cell entrapment systems, has been abandoned due to complexities in its operation. The operation of acoustic separation and hydrocyclones has been limited to small-scale bioreactors and more work needs to be done to utilize their full potential. A 200 L/day acoustic separation (BioSep) system is available and a 1000 L/day unit is being developed. Currently there are three systems that can be used at industrial scale. *Alternating tangential filters* (ATF), "*gravitational*" (replace with "*inclined settlers*"), and centrifuges can reach cell densities more than 20 million cells/mL, can be operated for a long time (>60 days) and at high perfusion rates (>1000 L/day). These systems are being optimized gradually and higher performance can be expected in the future in terms of cell density and achievable perfusion rates.

Control and Operation of Perfusion Bioreactors

Consistency and Reproducibility of Perfusion Cultures

Compared to batch or fed-batch systems, perfusion cultures generally result in a more uniform environment for the cells. Thus, better consistency and reproducibility is expected from a perfusion bioreactor. However, as discussed widely in the literature, this is not always the case and effective control strategies should be in place to obtain consistency.

The perfusion systems operate for a long time and if there are slight differences in operation, the results can vary from one bioreactor to another. Small differences and minor incidents can accumulate over time and result in completely different out-comes. The bioreactors were initiated using the same inoculum and operated under the same conditions. A few incidents including a problem in the feed in one of the bioreactors resulted in completely different performance over a 60-day run. Both cell densities and the product titer in bioreactor Run1 were lower for much of the time. This example demonstrates the need for a close monitoring and control of perfusion bioreactors.

Off-Line and On-Line Monitoring of Perfusion Bioreactors

Physical parameters (such as pH, dissolved oxygen, and temperature) in a perfusion bioreactor are typically monitored on-line and controlled in real time. Determination of cell density, viability, metabolite, and product concentration is performed using off-line sampling. The information obtained from these measurements is used to assess the culture performance and to make adjustments to the bioreactor if necessary. The frequency of sampling and making adjustments depends on the application. While a daily sampling is sufficient for most applications, more frequent and almost real-time monitoring and control are warranted for some processes.

On-line monitoring of perfusion cultures was demonstrated for cell density, oxygen consumption rate, metabolite concentrations, and for product. These measurements can be used to calculate cellular activities such as cell growth rates, cell specific metabolic, and production rates. In addition, these measurements can allow real-time control of perfusion bioreactors.

Operation of Perfusion Bioreactors by Dynamic Perfusion Rate Adjustments

In addition to physical parameters such as pH, dissolved oxygen, and temperature, perfusion rate and cell density are the important parameters to be monitored with perfusion bioreactors. Typically, the perfusion bioreactors start with a batch growth phase with no feeding. During that period, cells are allowed to grow to cell densities of 1–2 million cells/mL. Then the perfusion operation starts with continuous harvesting and feeding. The perfusion rate typically refers to the harvest flow rate, which is manually set to the desired value. A weight control for the bioreactor activates the feed pump so that a constant volume in the bioreactor can be maintained. Alternately a level control can be achieved by pumping out culture volume above a pre-determined level in a variable bleed stream.

The perfusion rate in the bioreactor must be adjusted to deliver sufficient nutrients to the cells. As the cell density increases in the bioreactor, the perfusion rate must be increased. The dynamic adjustment of the perfusion rate allows cells to grow and produce without nutrient depletion. There are several methods for perfusion rate adjustments and they are described in detail in the following sections.

Control of perfusion rate using cell density measurements

Cell density is the most important measurement used for perfusion rate adjustments. Depending on how the cell density measurements are conducted, perfusion rates can be adjusted daily or in real time. Several on-line probes have been developed for the estimation of cell density. Some of these cell density probes are robust and reliable, and can be used for automatic control of perfusion rates. These cell density probes can also be used to control the cell density at a desired set point by removing excess cells from the bioreactor.

Control of perfusion rate using oxygen consumption

Oxygen consumption or oxygen demand to the bioreactor is an indicator for cell density. These variables can then be used to adjust the perfusion rate. Oxygen consumption rates can be obtained using several techniques. Most of these techniques result in continuous on-line measurements. The perfusion rate can be adjusted proportionally to the oxygen consumption rate. The use of the oxygen consumption rate for perfusion rate control has been demonstrated successfully in laboratory and pilot plant scale bioreactors.

If the oxygen consumption rate is difficult to obtain, the gas flow rate to the bioreactor can also be used as an indicator for cell density. Thus, the perfusion rate can be adjusted based on the oxygen flow rate to the bioreactor.

Control of perfusion rate using metabolite measurements

On-line or off-line measurement of metabolites allows determination of metabolic rates of the cells through a mass balance in the system. This metabolic rate information can then be used to adjust the perfusion rate to the bioreactor. Ozturk et al. described the method of controlling glucose and lactate by dynamically adjusting the perfusion rate using a simple feed-forward algorithm.

Cell Specific Perfusion Rate Control

Ozturk et al. introduced the concept of *cell specific perfusion rate* (CSPR) for the design, optimization, and control of perfusion bioreactors using mass balances for substrate and products. The steady state mass balance equations for the substrates and products can be written as:

$$D \, . \, (S_o - S) \; q_s \, . \, X_v \text{ (for substrates)}$$

$$D \, . \, (p_o - p) \; q_p \, . \, X_v \text{ (for products)}$$

where c is the dilution rate; X_v is viable cell density; S, P are substrate and product concentrations in the reactor, respectively; S_o, P_o, are their concentrations in the feed; and q_s, q_p are the rates of substrate utilization and product formation.

Using CSPR, which is defined as CSPR = D/X_v, transforms these mass balance equations to:

S = $(S_o - q_s)$/CSPR (for substrates)

P = $P_o + q_p)$/CSPR (for products)

Therefore, if the CSPR is maintained as a constant, and the cellular activities do not change with time and cell density, then the CSPR control allows operation of high density reactors at constant medium composition, thus allowing consistent and optimal production.

The implementation of CSPR control to perfusion bioreactors was successfully demonstrated in pilot size bioreactors. A highly automated bioreactor control system monitors the cell density using optical density probe, determines the oxygen consumption rate in real time, and combines these measurements to estimate viable cell density and cellular activities. The system then controls the cell density and perfusion rate at set-points. A bioreactor with CSPR control could be run in an autopilot mode, make the necessary adjustments in response to the perturbations, and maintain a constant and uniform environment to the cells.

Optimization of Perfusion Bioreactors

Design and Operation of Perfusion Bioreactors at High Cell Densities

The bioreactors used for perfusion are not very different from those used for batch/fed-batch cultures, except that they are more compact in size and are connected to a cell retention device. The production rate in a 1000 L perfusion bioreactor can reach or even exceed the output of a 10,000 L fed-batch bioreactor.

Cell retention

The design and optimization of cell retention devices is key for the development of a perfusion process. The device should be robust, reliable, and scalable. There is no question that the addition of a cell retention device introduces complexity to the operation and that is why the industry is reluctant to accept perfusion processes for cell culture applications. In-house expertise is needed to design, operate, and scale-up these systems. There are already a variety of cell retention technologies that can be used at industrial scale. These technologies can be acquired easily and with some extra development and fine-tuning, these systems can be customized for a specific application. While the current systems can be developed further to accommodate higher perfusion rates, new technologies may emerge in the field of cell retention. Even though these developments are likely to be gradual, they will have a major impact on the perfusion bioreactors.

Aeration

High cell densities achieved in perfusion bioreactors pose unique challenges with respect to *aeration*, *mixing*, and *process control*. Use of sparging can deliver sufficient oxygen at high densities. However, foaming and cell lysis due to sparging require a careful sparger design and aeration strategies. Aeration also serves as a means to remove CO_2 from the bioreactors. Accumulation of CO_2 due to the differences in mass transfer coefficients can be a problem, as illustrated in the literature. This problem needs to be addressed especially in large-scale bioreactors. In addition to increased base addition, CO_2 accumulation can affect the cell growth and productivity and product quality.

Mixing and homogenization

Issues with mixing are elevated in perfusion bioreactors because of high cell densities. The mixing times in cell culture reactors can be as high as several minutes due to gentler agitation. At higher cell densities, viscosity in the bioreactor increases, thereby decreasing mixing efficiency. Poor mixing in bioreactors can result in gradients in pH and dissolved oxygen as discussed elsewhere.

In bioreactor mixing is used to suspend the cells, to disperse the bubbles for aeration, and to homogenize the added base. In poor mixing conditions, base addition, if not properly done, can cause localized cell lysis and impact the culture negatively. When time is not allowed for the homogenization of the base, hot spots with high pH are generated. Poor mixing also promotes heterogeneity of the culture and imposes control problems. Depending on the mixing and the eddy size distribution in the reactor, cells can aggregate to different sizes. Aggregation can induce segregation of the cells, with bigger ones migrating to the low agitation zones.

Process control

While a batch or fed-batch process can be operated with minimal automation, a high-level process control scheme is desirable for controlling the perfusion bioreactors. Perfusion bioreactors are more dynamic in nature because of two reasons: (i) the cell densities change over a wide range during the operation, and (ii) the system attains a high metabolic state due to the high cell densities involved. The bioreactors need to be monitored closely and adjustments need to be performed in a timely fashion. As mentioned before, several on-line monitoring and real-time control strategies have been developed for perfusion bioreactors. Some of these systems are highly automated and require complicated implementation strategies.

Scalability of perfusion bioreactors

Even though the perfusion bioreactors are more difficult to design and operate compared to batch or fed-batch counterparts, the scalability is rather straightforward. The success of scalability is mostly determined by the cell retention system and the control strategy used. Some of the cell retention systems such as spin filters are more difficult to scale-up. On the other hand, inclined plate settlers can be scaled-up easily using geometric similarities. A reliable control strategy, such as CSPR control, ensures the cellular environment to be similar at different scales and secures the scalability. As can be seen, the process is scaled-up successfully from laboratory to commercial scale.

New Directions in Perfusion Cultures: Medium Enrichment

The performance of a perfusion bioreactor is directly related to the cell concentration. The nutritional composition of the medium used, or the medium depth, is one of the key variables determining the maximum number of cells in the bioreactor. The other variable is the medium exchange rate in the bioreactor. Most of the current medium formulations can support about 10 million cells/mL (calculated based on cell yields on different nutrients) at growth conditions with an exchange rate of one volume per day. If the same medium is used for perfusion, the medium has to be exchanged 10 volumes per day to maintain a cell density of 100 million cells/mL. This magnitude of medium exchange, or perfusion rate, can impact the perfusion process significantly. First of all the cell retention system should still be functional at these high perfusion rates. Second, there will be an issue with medium utilization and cost. Third, the product from the bioreactor will be diluted at high perfusion rates.

Requirement for high medium exchange and relatively low product concentration from perfusion bioreactors have been the major disadvantages for perfusion processes. To deal with these issues several strategies have been developed.

A simple method for minimizing the medium utilization is to operate the bioreactor at low perfusion rates. If the cell density in the bioreactor can be maintained at low perfusion rates, and if the product is stable and the productivity of the cells is not impacted, the bioreactor can be operated at lower perfusion rates. This results in higher product concentration in the bioreactor without impacting the overall productivity. Initially the bioreactor is operated at a one volume per day exchange rate and a product concentration of 250 mg/L is obtained. A drop in perfusion rate from one volume per day to 0.5 volumes per day roughly doubles the concentration of product. Finally, a drop to 0.3 volumes per day results in product concentrations around 650 mg/L. While the results are encouraging and a higher product concentration can be obtained, it may be noted that the volumetric productivity in this case has decreased by 20%. A more effective solution for minimization of medium utilization and enhancing product concentration lies in the development of special concentrated or fortified medium for perfusion reactors. Research in the early 1990s demonstrated the use of 2-or 3-fold concentrated mediums for perfusion cultures. When the medium is concentrated and formulated at optimal osmolarity, the necessary perfusion rates are reduced by several fold. The use of enriched medium for perfusion has been shown to be successful. Accumulation of knowledge in cell metabolism and medium formulation over the years resulted in further developments of medium fortification.

The approach for medium enrichment is similar to fed-batch process development. If one analyzes the cell metabolism and the nutritional requirements of the cells, certain components can be added to the feed medium to reach higher cell densities. Instead of adding these components to the bioreactor as a feeding solution in the fed-batch case, these components can be added to the feed medium to the perfusion bioreactor resulting in an enriched medium.

The perfusion bioreactor starts with a base medium (Medium-1) and a cell density of about 15 million cells /mL is obtained. The medium is switched on Day 17 to Medium-2 (a richer medium), and to Medium-2 supplemented with *enriched medium* (EM) on Day 30. Due to medium enrichment during the course of this run, the viable cell density reaches 35 million cells/mL. The viability in the bioreactor remains more than 50% during the run. The product concentration increases by more than 100% and reaches 750 mg /L in enriched medium. The data demonstrate an interesting feature of enriched medium. A drop in glucose and a corresponding increase in lactate concentration are obtained in regular medium (Medium-1). This metabolic profile is very typical for CHO cells in culture. After the switch to enriched medium, a significant drop in lactate levels is observed. In fact the lactate concentration in the perfusion bioreactor reaches zero after the switch to enriched medium. These data show an efficient utilization of glucose in enriched medium.

These results demonstrate that there is great potential for the development and utilization of enriched medium for perfusion cultures. These developments are parallel to the feeding strategies for fed-batch processes. Understanding of cell metabolism and a careful formulation of enriched medium will definitely open new frontiers for perfusion bioreactors.

Concluding Remark

A number of mammalian cell retention methods are now available for achieving high cell density in a perfusion culture, with several of these methods scaled up successfully to industrial scale bioreactors. Culture productivity in high density perfusion cultures can be as high as that of fed-batch cultures while the culture duration can be extended significantly over that of the fed-batch cultures. The high medium exchange rate of the perfusion culture not only provides the necessary nutrient supply to the high concentration of cells in a bioreactor, but also removes the secreted products quickly from the bioreactor, thereby minimizing their exposure to degradative enzymes that accumulate in a fed-batch reactor. Consequently, the size of a perfusion bioreactor can be much smaller than that used for batch and fed-batch culture processes for the same amount of desired production. A large number of industrial perfusion processes are already approved by the regulatory agencies for production of therapeutic processes; previous concerns about getting such processes approved are no longer an issue. If a given mammalian cell line is sufficiently stable in its protein production, then perfusion cultures represent a superior bioreactor operating strategy for maximizing production of the protein.

8

Corneal Stem Cells

The cornea is a relatively simple organ, composed of three distinct tissue layers. One of these tissues is a self-renewing epithelium long believed to harbor a resident stem cell population. The other two corneal tissues are largely quiescent after infancy, and until recently they were not considered to undergo self-renewal or maintain resident stem cells. Over the last decade these views have changed. The location and characteristics of the corneal epithelial stem cells have now been described by a number of research groups, and populations of these cells have been expanded and used therapeutically. In addition, cell populations with characteristics of adult stem cells have been isolated and characterized from the stoma and endothelium of corneas. This chapter describes the methods used in identification, isolation, and culture of these three populations of cells. We also review data that speak to the stem cell character of these cells and their potential for use in therapeutic and tissue engineering applications.

Structure and Cells of the Cornea

The cornea is the window of the visual system. As the outermost layer of the eye, the cornea both serves as a barrier and provides the essential optical function of transmitting light to the retina. In addition to the transmission of light, the cornea provides 70–75% of the refractive power required to focus the light into an image. The cornea is made up of three unique differentiated cell types separated by basal laminae. The outermost layer consists of a stratified squamous, nonkeratinized epithelium. This tissue is supported by a basement membrane overlying an acellular zone of connective tissue known as the *Bowman layer*. The stroma, a collagenous connective tissue, makes

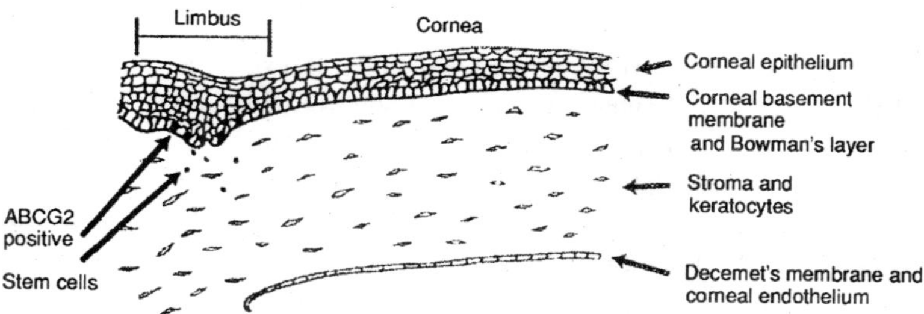

Fig. 8.1. Diagrammatic section through cornea.

up 90% of the cornea. It is populated with keratocytes, neural crest-derived mesenchymal cells that secrete the unique transparent tissue of the corneal stroma. The most posterior boundary of the cornea is the endothelium, comprised of a single layer of flattened cuboidal cells that maintain corneal transparency by regulating corneal hydration. Separating the corneal endothelium and the stroma is a basal lamina known as the *Descemet membrane*. Formation of the human cornea begins at approximately 5 to 6 weeks of gestation. After the lens vesicle separates from the surface ectoderm, the latter forms a layer of cuboidal epithelial cells, which develop into the corneal epithelium. Neural crest cells migrate between this epithelium and the lens, forming the *corneal endothelium*. A second wave of migration of cells from neural crest subsequently forms the stroma.

The three cellular layers of the cornea differ markedly in mitotic and self-renewal abilities. In the corneal epithelium, mitotically active basal cells continuously renew the nonmitotic population of suprabasal cells, which subsequently flatten as they migrate to the surface, where they are lost by desquamation. The stromal keratocytes, on the other hand, show little cell division in the normal adult. They undergo rapid cell division after localization in the cornea in late embryogenesis, but after birth the keratocyte cell number stabilizes and little or no mitosis can be detected throughout the lifetime. In the case of inflammation or wounding, however, the stromal keratocytes become activated and mitotic. The phenotype of the activated keratocytes changes to resemble that of fibroblasts and myofibroblasts, and connective tissue matrix secreted by these cells during wound-healing becomes opaque scars. After healing the cells become quiescent, but human corneal scars are very slow to resolve and it is not clear whether the resident cells return to a fully keratocytic phenotype. These properties suggest only a limited means of tissue renewal in the corneal stroma.

Renewal of corneal endothelial cells is even more limited than that of the keratocytes. After childhood, human corneal endothelial cells do not divide. Compensation for endothelial damage is accomplished by flattening of the remaining cells to cover the posterior surface of the cornea. In vitro, human corneal endothelial cells show only limited ability to divide after infancy.

These characteristics have led to the conventional view that the corneal epithelium is maintained by a stem cell population, but that the stroma and the endothelium, with limited ability for self renewal, are not products of tissue-resident stem cells.

Evidence for Stem Cells in the Cornea

Stem cells, by definition, undergo asymmetric cell division, that is, they undergo self-renewal while giving rise to differentiated daughter cells. Embryonic stem cells derived from the inner cell mass of the blastocyst are pluripotent, giving rise to most cells of the body. In culture, embryonic stem cells can be propagated indefinitely in an undifferentiated state. Historically, self-renewing adult tissues such as dermis have been thought to be generated by tissue-resident stem cells capable of generating only one type of cell, making them unipotent. The corneal epithelium is such a tissue, a rapidly regenerating stratified squamous layer covering the external surface of the cornea. The epithelium contains 5 layers of cells in the center and 10–11 layers in the transition zone between cornea and conjunctiva known as the *limbus*. Although the anatomic structure termed the limbal palisades of Vogt is implicated as the site responsible for corneal epithelial self-renewal, the exact location of the self-renewing progenitor cells remained obscure until Schermer and co-workers demonstrated that corneal epithelial stem cells reside in the limbal basal epithelium. This location was determined by analysis of the expression pattern of a major corneal epithelial keratin, K3, which is absent in a small population of these basal limbal epithelial cells. Injury or disease destroying the basal limbal cells causes a loss of corneal epithelium and its replacement with conjunctival epithelium, resulting in loss of corneal transparency. The best current marker for stem cells is a long-term retention of DNA-labeling by [^{3}H]thymidine or bromodeoxyuridine. This property is indicative of a slow-cycling population of cells and distinguishes the stem cells from the more rapidly cycling basal "*transient amplifying*" daughter cells, which are found centrally as well as in the limbus. Despite the strong evidence for the existence of limbal stem cells, methods have not been yet developed for isolation and culture of pure populations of these cells. Explants of limbal tissue containing presumptive stem cells can be expanded in culture and have been used in experimental animal models of limbal stem cell deficiency and clinically in patients with limbal stem cell deficiency to restore healthy corneal epithelial function. Such stem cell grafts are most successful with autologous tissue obtained from the contralateral eye of the same patient. Allografts usually fail without continuous antirejection therapy.

Over the past 10 years it has become apparent that stem cells in adult tissues are not restricted to self-renewing epithelia or hematopoietic tissues. Cells with properties of stem cells, including multipotential differentiation capability and extended life span, have been isolated from a number of adult mesenchymal tissues. The corneal stroma is a mesenchymal connective tissue making up 90% of the corneal thickness, with physical properties that provide the cornea its essential character. During development the stroma is produced by mesenchymal neural crest cells as they differentiate into keratocytes and begin to synthesize and secrete an extracellular matrix composed of collagens I, V, and VI and proteoglycans. As maturation proceeds, the stroma dehydrates, becomes thin and transparent, and contains flattened and interconnected keratocytes. In late embryogenesis, chicken keratocytes appear to retain neural crest progenitor properties after transplantation into a new environment along cranial neural crest migratory passageways.

In adult mammals, however, numerous in vitro experiments show that keratocytes rapidly lose their characteristic phenotype after several population doublings. Such a loss of phenotype occurs in healing wounds in vivo as well. Recently, the authors found that the stroma of bovine and human corneas contain small populations of cells exhibiting self-renewal ability for an extended number of population doublings in culture. These stromal progenitor cells demonstrated potential for differentiation into several noncorneal cell types, a characteristic similar to that of adult stem cells from other mesenchymal tissues. These corneal stromal stem cells can be cloned and proliferate in vitro for more than 100 doublings. Currently the function of the stromal stem cells in vivo, especially during wound healing, is unclear. Some researchers have suggested that in corneal stroma there are stem cells derived from bone marrowthat are CD34 positive. Nakamura et al. [2005] found that lethally irradiated mice, rescued by tail vein injection of bone marrow cells from *green fluorescent protein* (GFP)-expressing mice, exhibited a resident population of green cells in the cornea. The function of bone marrow-derived stem cells in cornea and the relationship between bone marrow-derived cells and keratocytes remain unknown.

The corneal endothelium is a single layer of flat hexagonal cells forming a boundary between the corneal stroma and anterior chamber and functioning as a pump to regulate stromal hydration. The endothelium permits the passage of nutrients from the aqueous humor into the avascular cornea through a leaky barrier formed by focal tight junctions. The endothelium simultaneously removes water and CO_2 from the stroma via the activity of Na^+/K^+-ATPase and bicarbonate-dependent ionic pump. The pump protein is located mainly on the lateral plasma membranes. Although, like the keratocytes, the corneal endothelium is derived from neural crest, the endothelial cell characteristics are different from those of keratocytes.

The human corneal endothelium comprises a monolayer of polygonal cells, which are arrested in G_1 phase in vivo and do not normally replicate to replace dead or injured cells. This relative lack of cell division results in a physiological reduction of cell density of about 0.3–0.5% per year. Scattered evidence, however, suggests the potential for some mitotic events in human corneal endothelium. Mitotic figures were observed in vivo by specular microscopy after rejection reaction of a corneal graft. Clusters of cells smaller than surrounding cells suggested that at least under some circumstances mitosis occurs in the endothelium of the adult human cornea. Recently Yokoo et al. identified cells in the human corneal endothelium that can form cell spheres in attachment-independent culture. Cells in these spheres, which form under conditions similar to those used for isolation of neural stem cells, can be expanded and can generate daughter cells expressing neuronal and mesenchymal molecular markers. These properties suggest a stem cell origin for the cells forming the spheres.

The sphere-forming cells also adapted the polygonal morphology characteristic of endothelial cells, suggesting that these cells are endothelial progenitors. Mimura et al. showed that these precursors were effective in vivo in restoring endothelial function in an animal model of corneal endothelial deficiency. The same group observed peripheral and central rabbit corneal epithelia to contain a significant number of precursors, but the peripheral endothelium contains more precursors and has a stronger self-renewal capacity than the central region by sphere-forming assay. Because the long-term culture of human endothelial cells has not been carried out and because of the lack of both stem cell and endothelial markers, positive identification of the proposed endothelial stem cells in situ has yet to be carried out.

Identification of Corneal Stem Cells

Although the concept that corneal epithelial stem cells reside in the limbus is widely accepted, identification of these cells has been difficult because of the lack of unique molecular markers. Label retention has been the most robust means of stem cell identification in experimental animals, but it has not been feasible for human epithelial stem cells. One candidate marker, p63 protein, a transcription factor homolog of the tumor suppressor p53, is highly expressed in the basal or progenitor layers of many epithelial tissues and is essential for regenerative proliferation. Whether p63 is a marker of limbal stem cells, however, is still controversial. Pellegrini et al., [1999, 2001] investigated p63 as a specific marker of keratinocyte stem cells based on its localization to a subset of basal limbal cells and clonal analysis showing preferential p63 expression in colonies with the greatest clonogenic potential.

Wang et al. [2003] suggested that p63 is not a stem cell-specific marker, based on its coexpression with Ki-67 (a proliferating cell marker) in suprabasal nuclei of rabbit corneal limbal sections and limbal explants on amniotic membrane. Moore et al. [2002] and Joseph et al. [2004] described the p63 expression pattern of mouse and rat keratoepithelial grafts and human limbal explants, respectively, both noting that p63 is not exclusive to the stem cells, based on the continued p63 expression of cells migrating out from cultured explants and its expression in sections from the central cornea. Based on the coexpression of connexin-43 (a negative marker for limbal stem cells) and p63 in a monolayer of cultured human corneal limbal cells formed after 2–3 weeks of culture, Du et al. [2003] concluded that p63 is a marker of stem cells as well as *transient amplifying* (TA) cells in limbal cultures. Chen et al. [2004] identified p63 as one of a few markers that could define a putative stem cell phenotype based on its differential expression in the basal cells of human limbal sections, bringing the controversy of the utility of p63 as a corneal epithelial stem cell marker to the foreground.

More recently, Salehi-Had et al. [2005] suggested that p63 expression in culture cannot be used as a marker for stem cells, based on the observation that the majority of corneal limbal epithelial cells express p63 in colonies derived from single cells and in subconfluent cultures regardless of time in culture or continuance of cell division.

Keratins 3 and 12 (K3 and K12) are widely used to identify differentiated corneal epithelial cells, and connexin-43 is expressed in basal epithelial (TA) cells but not basal limbal (stem) cells. Thus these antigens currently serve as negative markers to identify the differentiated epithelial cells.

The most recent and currently best candidate for a marker of corneal epithelial stem cells is a drug-resistance transporter protein known as ABCG2. Adult stem cells have the ability to efflux fluorescent dye Hoechst 33342, leading to reduced red and blue fluorescence in *fluorescence-activated cell sorting* (FACS). These cells are referred to as "*side population*" (SP) cells because in the two-dimensional display of red and blue fluorescence, cells having reduced Hoechst dye appear as a small tail to the left side of the mass representing live somatic cells. SP cells from a number of adult tissues have been shown to exhibit many characteristics of stem cells.

The SP cells are lost after treatment with verapamil, a drug that blocks action of the ATP-binding cassette transporter G family member ABCG2. This transporter protein has been identified as the Hoechst efflux pump and as a specific marker for many kinds of stem cells such as hematopoietic, mesenchymal, muscle, neural, cardiac, islet, and keratinocyte stem cells. Recent studies have shown the presence of side population cells present in corneal epithelial and stromal tissue. In these cells, ABCG2 protein and mRNA expression has been found to be correlated with the SP phenotype and with stem cell characteristics. The ABCG2-positive cells of cornea are located in the limbus. In the epithelium these cells are localized in unique crypts associated with the palisades of Vogt. In the stroma ABCG2-staining cells are seen just posterior to the limbal basement membrane.

Preparatino of Media and Reagents

PBSA/GASP

PBSA containing antibiotics: gentamicin, 50 μg/mL; amphotericin B, 1.25 μg/mL; streptomycin, 100 μg/mL; penicillin, 100 U/mL (GASP)

CMF-Saline G

1. NaCl	8 g/L
2. KCl	0.4 g/L
3. $Na_2HPO_4 \cdot 7H_2O$	0.29 g/L
4. KH_2PO_4	0.15 g/L
5. Glucose	1.1 g/L
6. pH	7.2

CMF/GASP

CMF-Saline G with GASP antibiotics

Trypsin/EDTA

Trypsin 0.25%, Na_2EDTA 0.5 mM, in CMF-Saline G

DMEM/F-12/GASP

DMEM/F-12 with GASP antibiotics

LSC Culture Medium

DMEM/F-12 containing:

1. FBS	5%
2. Human epidermal growth factor (EGF)	20 ng/mL
3. Glutamine	4 mM
4. Triiodothyronine	2 nM
5. ITS	
Insulin	5 μg/mL
Transferrin	5 μg/mL
Selenous acid	5 ng/mL
6. Hydrocortisone	0.5 μg/mL
7. Cholera toxin	30 ng/mL
8. Adenine	0.18 mM
9. GASP antibiotics	

DMEM/10FB/GASP

DMEM containing 10% FBS and GASP antibiotics

Modified Jiang Medium (MJM) for Culture of Human Stromal Stem Cells

DMEM/MCDB-201, 60:40, with:

1. Fetal bovine serum (FBS)	2%
2. Epidermal growth factor (EGF)	10 ng/mL
3. Platelet-derived growth factor (PDGF-BB)	10 ng/mL
4. ITS	
5. Leukemia inhibitory factor (LIF)	200 U/mL
6. Linoleic acid–bovine serum albumin (LA-BSA),	1 mg/mL
7. Ascorbic acid-2-phosphate	0.1 mM
8. Dexamethasone	1×10^{-8} M
9. GASP antibiotics	

HBSS/2FB

Hanks' BSS with 2% FBS

DMEM/2FB

DMEM with 2% FBS

Keratocyte Differentiation Medium

Advanced D-MEM (Invitrogen) with:

1. Basic fibroblast growth factor (FGF-2)	10 ng/mL
2. Ascorbic acid-2-phosphate	0.1 mM

Chondrocyte Differentiation Medium (CDM)

DMEM/MCDB-201, 60:40 with:

1. FBS	2%
2. Ascorbic acid-2-phosphate	0.1 mM
3. Dexamethasone	1×10^{-7} M
4. TGF-β	10 ng/mL
5. Sodium pyruvate	100 μg/mL

Neural Differentiation Medium (NDM)

Advanced D-MEM with:

1. Epidermal growth factor (EGF)	10 ng/mL
2. FGF-2	10 ng/mL
3. All-*trans* retinoic acid	1 μM

Corneal Endothelial Cell Medium (CECM)

DMEM/F-12, 1:1, with:

1. FBS	2%
2. FGF-2	10 ng/mL
3. EGF	10 ng/mL
4. PDGF-BB	10 ng/mL
5. ITS, 1×	
6. Ascorbic acid-2-phosphate	0.1 mM
7. GASP antibiotics	

Endothelial Spheres Medium (ESM)

DMEM/F-12, 1:1, with:

1. Methylcellulose	1.5%
2. B27 supplement, 1×	
3. FGF-2	10 ng/mL
4. EGF	20 ng/mL
5. GASP antibiotics	

SDS Sample Buffer (6×)

1. Tris·HCl, 0.5 M, pH 6.8	7 mL

2. Glycerol 3 mL
3. SDS 1 g
4. Bromophenol blue 1.2 mg

Concentrations in 6× stock: 0.35 M Tris, 30% glycerol (v/v), 10% (w/v) SDS, 0.012% (w/v) Bromophenol Blue.

Use at final concentrations: 0.058 M Tris, 5% glycerol (v/v), 1.67% (w/v) SDS, 0.002% (w/v) Bromophenol Blue.

Blocking Buffer

PBSA + 0.5% BSA + 2% normal goat serum

Culture of Human Corneal Limbal Stem Cells (LSC)

Although side population cells can be identified and isolated from the limbal epithelium by FACS, this purification method has not been adopted widely as a means for isolation of pure populations of viable LSC for culture. Typically, unfractionated populations of human corneal limbal cells are cultured and passaged in vitro under conditions favoring stem cell expansion. These cultures have been transplanted in vivo for treatment of limbal stem cell deficiency. For in vitro culture of LSC, there are three methods: (i) culture directly on plastic, (ii) culture on amniotic membrane, and (iii) culture with a feeder layer. The authors compared culture procedures on plastic and amniotic membrane and concluded that culture on human amniotic membrane suppresses differentiation of limbal progenitor cells and promotes their proliferation. Tseng and co-workers concluded that the coculture system with a feeder layer of mitomycin C-treated mouse 3T3 cells promoted more clonal growth of limbal progenitor cells. We introduce these different culture systems below.

Safety note: Human tissues should be handled in Biosafety Level 2 laboratories, using a laminar flow hood and protective personal apparel as approved by the institutional review board as appropriate for this purpose. Serology should be obtained from each donor to exclude possibility of contamination with HIV and hepatitis.

Preparation of Substrata

Protocol - 1. Preparation of human amniotic membrane (HAM)

Reagents and materials

Sterile or aseptically prepared

1. Human placenta. *Note:* Human tissue for research must be obtained with informed consent of the donor, using a protocol approved by the institutional review board. Preservation of the anonymity of

donors, procedures in accordance with the Declaration of Helsinki, and biosafety concerns will need to be addressed.

2. PBSA/GASP
3. Glycerol
4. DMEM containing 50% glycerol
5. Trypsin/EDTA
6. Millicell microporous membrane tissue culture insert
7. Plastic spatula and sterile cotton swab

Procedure

1. Wash the tissue with PBSA/GASP.
2. Separate the HAM, a thin sheet consisting of the epithelium, basement membrane, and some underlying compact stroma, manually, using a gloved hand to slide under the HAM and separate it from the stromal tissue of the placenta.
3. Use the sterile cotton swabs to remove the underlying compact stroma from the basement membrane to keep the amniotic membrane as thin as possible.
4. The crude HAM can be stored in DMEM containing 50% glycerol at –70°C, pending testing of donor sera for disease.
5. Immediately before use, thaw the HAM, wash it with PBSA, and cut it into pieces approximately 2 cm in diameter.
6. Digest the pieces with Trypsin/EDTA at 37°C for 30 min.
7. Scrub the digested HAM gently with a plastic spatula to remove the epithelium without breaking the basement membrane.
8. Wash the denuded membrane with PBSA and allow it to adhere onto a Millicell microporous membrane tissue culture insert with the basement membrane side (from which epithelial cells have been removed) facing up.

Protocol - 2. Preparation of mouse NIH 3T3 fibroblast feeder layer

Reagents and materials

Sterile or aseptically prepared

1. Mouse NIH 3T3 fibroblasts
2. DMEM/10FB/GASP
3. Trypsin/EDTA
4. Mitomycin C (MMC), 500 μg/mL in water
5. Tissue culture-treated plastic flasks, 25 or 75 cm^2
6. Tissue culture plates, 6-well

Nonsterile

1. Optional (alternative to mitomycin C treatment): gamma or X-ray irradiator
2. Low-speed refrigerated centrifuge

Procedure

1. Seed NIH 3T3 fibroblasts in 25-cm^2 or 75-cm^2 tissue culture flasks in DMEM10FB/GASP).
2. When the cells just reach confluence, add MMC (5 μg/mL) for 2 h at 37°C.
3. Trypsinize and plate the cells at a density of 2×10^4 cells/cm^2 in 3.5-cm dishes or six-well plates.
4. As an alternative to MMC, confluent cultures of 3T3 cells may be irradiated with 60 Gy, using ^{60}Co or an X-ray source. The cells are amitotic but still living and can be used as a feeder layer.

Protocol - 3. Harvesting corneal epithelium and stem cell isolation

Reagents and materials

Sterile or aseptically prepared

1. Whole human cornea
2. CMF-Saline G
3. Dispase II, 1.2 U/mL (2.4 U Dispase II diluted in DMEM/F-12/GASP)
4. Trypsin/EDTA
5. DMEM/F-12/GASP
6. DMEM/F-12/2FB/GASP: DMEM/F-12/GASP with 2% FBS
7. CMF/GASP
8. LSC culture medium
9. Curved iris scissors, 11 cm (4-3/8 in.)
10. Jeweler's forceps, 10 cm (4 in.)
11. Corneal scissors, 19-mm blades, sharp tip
12. Colibri suturing forceps, 0.1 mm

Nonsterile

1. Variable-speed rocking mixer
2. Dissecting microscope
3. Low-speed refrigerated centrifuge
4. Hemocytometer

Procedure

1. Wash the cornea 3 × 5 min in CMF-Saline G
2. Trim off the residual sclera, conjunctiva, and iris.
3. Add 2 ml of Dispase II and leave at 4°C, overnight, with gentle agitation.
4. Rock the cornea in Dispase II at 4°C for 30 min in a rocking mixer.
5. Wash the cornea once in DMEM/F-12/GASP.
6. Under a dissecting microscope, carefully remove the central corneal epithelium and peel off the limbal pigmented epithelium, which contains the palisades of Vogt.
7. Digest the limbal epithelial sheets in Trypsin/EDTA for 10 min at 37°C.
8. Add one volume of DMEM/F-12/2FB/GASP.
9. Centrifuge at 400 *g* for 10 min and discard the supernate.
10. Wash once with DMEM/F-12/2FB/GASP, centrifuge, and discard the supernate.
11. Add 1 mL of LSC culture medium, disperse the cell pellet, and determine the cell number with a hemocytometer.
12. Seed the cells at 1 × 10^4/ cm^2 into a tissue culture plate with a 3T3 feeder layer or on prepared HAM.
13. Change the medium every 3 days.
14. When the cells are 90% confluent, passage by trypsinization:
    (a) Remove the medium.
    (b) Wash 1× with CMF-Saline G.
    (c) Add trypsin/EDTA for 10 min at 37°C.
    (d) Whenthe cells are released,addanequalvolumeofDMEM/F-12/2FB/GASP.
    (e) Count the cells.
    (f) Centrifuge 400 *g* for 10 min.
    (g) Discard the supernate and add sufficient new medium to seed at a density of 1 × 10^4 cells/cm^2.
15. Alternatively, primary cells may be frozen in cryopreservation medium.

LSC Characterization and Differentiation

As mentioned above, LSC express ABCG2 and p63 but do not express connexin-43 and keratins 3 and 12. Thus these cells can be identified by immunostaining, RT-PCR, and immunoblotting.

LSC differentiation

LSC can differentiate into stratified epithelial layers after culture at an air-liquid interface. Passage the cells on a Millipore Millicell culture insert, submerged in LSC medium. When confluent, reduce the volume of culture medium in the well so that the layer of cells is bathed from the bottom by medium but the top is covered only by a thin film of medium. The level of medium must be checked daily. After 10–14 days, the cells will form stratified epithelial layers. These can be identified by standard histological staining and by immunostaining for connexin-43 and keratin 3.

Protocol - 4. Immunostaining limbal stem cell cultures

Reagents and materials

Nonsterile

1. Primary antibodies:

(a) ABCG2	mouse monoclonal BXP-21	1:100
(b) Connexin-43	rabbit polyclonal Cx43	1:100
(c) Cytokeratin 3 (K3)	mouse monoclonal AE5	1:500
(d) Cytokeratin 12 (K12)	rabbit polyclonal J7	1:200

2. Secondary antibodies:

3. Anti-mouse IgG	Alexa-488	1:2500
4. Anti-rabbit IgG	Alexa-546	1:1500

5. Blocking antibody: 10% heat-inactivated goat serum in PBSA
6. Triton X-100
7. PBSA
8. Paraformaldehyde, freshly made from 16% stock, 4% in PBSA.
9. Immu-Mount antifade mounting solution
10. Slides and #1 coverslips
11. Confocal microscope

Procedure

1. Culture cells either on amniotic membrane or on plastic.
2. Wash with CMF-Saline G once.
3. Fix in freshly made 4% PFA in PBSA for 15 min at room temperature.
4. Wash once in PBSA.
5. Permeabilize with 0.1% Triton X-100 in PBSA for 10 min.
6. Block nonspecific binding with 10% heat-inactivated goat serum in PBSA

7. Incubate samples with primary antibodies for 1 h at room temperature.
8. Wash twice with PBSA.
9. Add secondary antibodies and incubate for 1 h at room temperature.
10. Wash the samples twice in PBSA.
11. Mount in a minimal volume of antifade mounting medium under a #1 coverslip.
12. Photograph the samples with an epifluorescence or confocal microscope using a 40× oil objective.

Protocol - 5. Characterization of limbal stem cells with reverse transcription-PCR

Reagents and materials

1. Primers:

ΔN p63	Forward: CAGACTCAATTTAGTGAG
	Reverse: AGCTCATGGTTGGGGCAC
K12	Forward: CTA CCT GGA TAA GGT GCG AGC T
	Reverse: TCT CGC ATT GTC AAT CTG CA
β-Actin	Forward: GAG GCG TAC AGG GAT AGC AC
	Reverse: GTG GGC ATG GGT CAG AAG

2. RNA extraction kit, e.g., RNeasy (Qiagen)
3. Superscript II (Invitrogen)
4. DNase I (Ambion)
5. HotStarTaq (Qiagen)
6. Materials for 6% acrylamide gel electrophoresis
7. SYBR Gold (Invitrogen)

Procedure

1. Extract RNA from stem cells from one 3.5-cm dish by a standardized procedure, such as RNeasy.
2. Transcribe cDNA from 400 ng of total RNA, using random hexamers and Superscript II followed by DNase I treatment.
3. Carry out PCR on cDNA from 20 ng of RNA, using HotStarTaq. Cycling conditions: 94°C for 2 min, 94°C for 30 s, 55°C for 45 s, 72°C for 45 s, for 35 cycles, 72°C for 10 min.
4. Separate PCR products on 6% acrylamide gel and detect by SYBR Gold.

Immunoblotting

Connexin-43 and keratins 3 and 12 are detected by immunoblotting of proteins from cell lysates with procedures described above. ABCG2 can be immune precipitated from cell lysates after cell surface biotinylation as described below.

Cryopreservation

Briefly, for LSC cells, trypsinize, count, pellet by centrifugation, and resuspend the cells in freezing medium at 2–5 × 10^6 cells/mL with 1 mL per vial. Chill at a controlled rate of 1°C/h, using a commercial freezing box filled with isopropanol in a –80°C freezer overnight. On the next day, transfer the vials to liquid nitrogen. Freezing medium: 70% culture medium (DMEM/F-12), 20% FBS, 10% DMSO. Make fresh each time before freezing cells.

Variations and Applications of the Method

LSC can be cultured as limbal explants or purified by FACS for side population cells.

Explant culture

After Dispase II digestion of human cornea, isolate the limbal epithelium and cut it into pieces about 1 mm in diameter. Seed these pieces on 3T3 feeder layersor on amniotic membranewith the epithelial side up. The cells will migrate out from the tissue over a period of 2–4 weeks.

FACS isolation of limbal SP cells

This procedure is identical to that used for isolation of stroma side population cells by cell sorting. For epithelial cells, both primary cells and passaged cells can be sorted if enough cells can be obtained after tryptic digestion.

Culture of Human Corneal Stromal Stem Cells

Corneal stromal cells, known as *keratocytes*, have a quiescent phenotype characterized by a unique dendritic morphology and a very low to nonexistent rate of proliferation *in vivo*. In response to acute injury, keratocytes become mitotic, adopt a fibroblastic phenotype, and move to the injured area. In vitro, primary keratocytes can be maintained in serum-free or low mitogen serum-containing culture medium in a quiescent state, exhibiting cellular morphology and matrix secretion similar to those of keratocytes in vivo. Fetal bovine serum can induce keratocytes to proliferate but also causes keratocytes to become fibroblasts and myofibroblasts.

Isolation of Stromal Cells

The isolating and culturing of stromal stem cells provides an important source of keratocytes for in vitro and in vivo research. We have demonstrated a method for obtaining human corneal stromal stem cells using FACS to isolate the SP cell population from primary stromal cell cultures. The isolated stem cells express ABCG2 and grow in vitro for more than 100 population doublings without loss of the ability to differentiate into keratocytes. Like quiescent primary keratocytes, the cells secrete the keratan sulfate proteoglycans lumican, keratocan, and mimecan, often identified as molecular markers for keratocytes.

Protocol - 6. Isolation of primary human corneal stromal cells

Reagents and materials

Sterile or aseptically prepared

1. Whole human cornea
2. CMF-Saline G
3. Dispase II
4. Collagenase type L, 1 mg/mL in DMEM/F-12/GASP
5. TrypLE Express or 0.25% trypsin in CMF-Saline G
6. DMEM//F-12/GASP
7. DMEM/F-12/2FB/GASP: DMEM/F-12/GASP with 2% FBS
8. CMF/GASP
9. Tissue culture-grade plastic dishes, 3.5 cm, or 6-well plates
10. Scalpel or single-edge razor blades
11. Cell strainer, 70-μm nylon mesh
12. Plastic spatula, "*Cell Lifter*" or "*Cell Scraper*"
13. Curved iris scissors, 11 cm (4-3/8 in.)
14. Jeweler's forceps, 10 cm (4 in.)
15. Corneal scissors, 19-mm blades, sharp tip
16. Colibri suturing forceps, 0.1 mm

Nonsterile

1. SDS sample buffer. Use at final concentrations: 0.058 M
2. Tris, 5% glycerol, 1.67% SDS, 0.002% bromophenol blue
3. Variable-speed rocking mixer
4. Tube roller apparatus
5. Dissection microscope
6. Centrifuge, low speed, refrigerated

Procedure

1. Wash the cornea 3 × 5 min in CMF-Saline G.
2. Trim off the residual sclera, conjunctiva, and iris.
3. Add 2 mL of 1.2 U Dispase II (2.4 U Dispase II diluted in DMEM/F-12/GASP) 4°C, overnight on a rocking mixer.

4. Rotate the cornea in Dispase II for 30 min at 4°C.
5. Wash the cornea once in DMEM/F-12/GASP.
6. Under the dissection microscope, carefully peel off the epithelium and endothelium.
7. Use a plastic spatula to scrape both epithelial side and endothelial side of the stroma. Observe through microscope to make sure all of these cellular layers are removed.
8. Wash the corneal stroma in new medium once.
9. Mince the stroma into 2-mm cubes, using scalpel, razor blades, or fine scissors.
10. Digest up to 3 h at 37°C in collagenase in DMEM/F-12/GASP until most of the tissues disperse.
11. Centrifuge at 400*g* for 10 min and discard the supernate.
12. Resuspend the cells in fresh DMEM/F-12/GASP, filter the digest through a 70-μm Cell Strainer, and repeat the centrifugation.
13. Repeat this wash a second time. Count the cell number after each spin.
14. Resuspend the primary stromal cells in MJM and seed into tissue culture coated plastic dishes at a density of 1×10^4 cells/cm^2.
15. Change the medium every 3 days.
16. When the cells are 90% confluent, passage by trypsinization.
    (a) Aspirate the medium.
    (b) Wash in CMF-Saline G.
    (c) Add trypsin or TrypLE to barely cover cells for 10 min at 37°C.
    (d) Add DMEM/F-12/2FB to terminate the digestion.
    (e) Count the cell number.
    (f) Centrifuge resuspended cells and discard the supernate.
    (g) Resuspend in sufficient fresh MJM to seed cultures at a density of 1×10^4 cells/cm^2.

Isolation of Stromal SP Cells with FACS

Protocol - 7. Selection of corneal stromal stem cells by hoechst 33342 efflux

Reagents and materials

Sterile or aseptically prepared

1. Stromal cells at passages 2–4
2. HBSS/2FB

3. Hoechst 33342 dye, 1 mg/mL in water
4. Propidium iodide, 2 mg/mL in water
5. Verapamil, 500 μg/mL in water
6. DMEM/2FB
7. Trypsin/TrypLE: TrypLE Express or 0.25% trypsin in CMF-Saline G
8. MoFlo (or similar) high-speed cell sorter, with 350-nm excitation

Procedure

1. Trypsinize passage 2–4 stromal cells.
2. Count the cell number and dilute to 1.0×10^6 cells/mL in DMEM/2FB.
3. Add 5 μg/mL Hoechst 33342 dye for 90 min at 37°C; agitate every 20 min.
4. As a control, some cells are preincubated for 20 min with 50 μg/mL verapamil before Hoechst 33342 incubation.
5. After staining, wash the cells twice by centrifugation in HBSS/2FB at 4°C and then resuspend and store them in cold HBSS/2FB on ice.
6. Immediately before sorting, add 2 μg/mL propidium iodide to identify nonviable cells.
7. Sort cells on a sterile, high-speed cell sorter, using 350-nm excitation. Collect the cells showing reduced fluorescence of both blue (670 nm) and red (450 nm). This "*side population*" is collected separately from dead cells and from fully labeled cells. As a control, confirm that the side population is eliminated by verapamil preincubation.

Alternatively, cells can be sorted according to expression of ABCG2 protein, although this procedure may not yield a population with the same level of "*stemness*" as side population cells.

Protocol - 8. Immunoselection of corneal stromal stem cells by ABCG2 expression

Reagents and materials

Sterile

1. Stromal cells at passages 2–4
2. PBSA/BSA: PBSA + 0.5% BSA
3. PBSA/2FB: PBSA with 2% FBS
4. Blocking buffer

5. Antibody: MAB 4155F, Clone 5D3 anti-ABCG2-FITC, or Isotype-FITC

Nonsterile

1. High-speed cell sorter with a 488-nm argon laser and a band-pass filter of 525/20

Procedure

1. Trypsinize passage 2–4 cells and count the cell number.
2. Spin down at 400 × *g*, 10min.
3. Wash once with PBSA/BSA.
4. Block in 50 μL of blocking buffer for 10 min on ice.
5. Add 10 μL of antibody (MAB 4155F, Clone 5D3 anti-ABCG2-FITC, or Isotype-FITC), 30 min, on ice.
6. Wash once by centrifugation in PBSA/BSA.
7. Gently resuspend the cells in 1 mL of PBSA/2FB. Keep the cells on ice until flow cytometry is performed.
8. Perform *fluorescence-activated cell sorting* (FACS), using a high-pressure, high-speed cell sorter. A 488-nm argon laser is used to excite the fluorescein isothiocyanate, and a band-pass filter of 525/20 is used to measure emitted light. Gates in the right angle scatter versus forward scatter diagrams are used to exclude debris. Collect at least 100,000 events for analysis.
9. The sorted ABCG2-positive and -negative cell populations can be used to evaluate their colony-forming efficiency and gene expression of stem cell markers and to passage for further investigation and in vivo transplantation.

Characterization and Differentiation

Stem cell characterization

Stromal stem cells in MJM will exhibit expression of PAX6 and ABCG2, genes not expressed by differentiated keratocytes.

PAX6

Stem cells in sparse conditions in MJM will express high levels of nuclear PAX6. Cells seeded at 1 × 10^4/cm^2 in MJM are fixed in 4% PFA in PBSA for 10 min, permeabilized in 0.1% Triton X-100 for 10 min, and blocked in 10% goat serum as described above. The cells are stained with anti-PAX6, diluted 1:100, and counterstained with cytoplasmic myosin (CMII25) followed by secondary antibodies of Alexa-546 anti-rabbit and Alexa-488 anti-mouse for 1 h. The cells are photographed by epifluorescence microscopy as described above.

ABCG2

ABCG2 can be detected by immune precipitation after cell surface biotinylation. Cell layers are rinsed in PBSA and incubated with Sulfo-NHS-LC-Biotin at 1 mg/mL in PBSA for 15 min on ice. Cell layers are washed again in PBSA, and the cells are scraped with a cell lifter and pelleted. Cells are lysed in 0.5 mL of M-PER and cleared with 10 μL of protein G magnetic beads. ABCG2 antibody (clone BXP-21) is preincubated with protein G beads, and then the loaded beads are incubated with samples overnight. The beads are collected and rinsed in PBSA, and the bound protein is eluted by heating in SDS sample buffer. Proteins are separated on 4%–20% SDS-PAGE gel and transferred to *polyvinylidene difluoride* (PVDF) membranes, and biotinylated protein is detected with streptavidin-horseradish peroxidase, using a luminescent substrate.

Keratocyte differentiation

As the stem cells differentiate to keratocytes they will lose expression of PAX6 and ABCG2 and express molecular markers unique to keratocytes. The most reliable of these markers are the proteoglycan keratocan and the glycosaminoglycan keratan sulfate. These both can be detected by immunostaining and Western blot. Keratocan mRNA can be quantified as well.

To induce differentiation, passage the stem cells at 1×10^4 cells/cm^2 in keratocyte differentiation medium. Change the medium every 2–3 days. After 1–2 weeks, the cells will be induced into keratocytes.

Immunodetection of proteoglycans

Proteoglycans are recovered from culture media by passage over ion exchange columns (SPEC-NH2 microcolumns). These are rinsed in 0.2 M NaCl, 6 M urea, 0.02 M Tris, pH 7.4, and eluted in 0.5 mL of 4 M guanidine-HCl, 0.02 M Tris, pH 7.4. The samples are dialyzed against water and lyophilized. Samples are resuspended in 100 μL of 0.1 M ammonium acetate, pH 6.5, and divided into half. One half is digested overnight at 37°C with 2 mU/mL Keratanase II and 2 mU/mL endo-β-galactosidase.

Digested and undigested samples, normalized for cell number, are run on a 4%–20% SDS-PAGE gradient gel, transferred to PVDF membrane, and subjected to immunoblotting with Kera-C polyclonal antibody against keratocan and monoclonal antibody J36 against keratan sulfate. The digested samples will not react with J36 but will show a sharp band of 55 kDa for the keratocan.

Quantitative reverse transcription–polymerase chain reaction (RT-PCR)

Quantitative RT-PCR is performed with SYBR Green RT-PCR Reagents according to the manufacturer's instructions. The reaction is carried out for 40 cycles of 15 seconds at 95°C and 1 minute at 60°C after initial incubation at 95°C for 10 minutes. Reaction volume is 50 μL, containing 1× SYBR Green PCR buffer, 3 mM Mg^{2+}, 200 μM dATP, dCTP, and dGTP and 400 μM dUTP, 0.025 units/mL AmpliTaq Gold polymerase, 5 μL cDNA, and forward and reverse primers at optimized concentrations. A dissociation curve for each reaction is generated on the Gene-Amp ABI Prism 7700 Sequence Detection System to confirm the absence of nonspecific amplification. Amplification of 18S rRNA is performed for each cDNA (in triplicate) for normalization of RNA content. Threshold cycle number (C_t) of amplification in each sample is determined by ABI Prism Sequence Detection System software. Relative mRNA abundance is calculated as the C_t for amplification of a gene-specific cDNA minus the average C_t for 18S expressed as a power of 2, that is, $2^{-\Delta Ct}$. Three individual gene-specific values thus calculated are averaged to obtain means ± SD.

Table 8.1. Human primers for detection of expression of keratocyte differentiation genes

Keratocan	Forward: ATCTGCAGCACCTTCACCTT
	Reverse: CATTGGAATTGGTGGTTTGA
ABCG2	Forward: TGCAACATGTACTGGCGAAGA
	Reverse: TCTTCCACAAGCCCCAGG
Pax6	Forward: CAATCAAAACGTGTCCAACG
	Reverse: TAGCCAGGTTGCGAAGAACT
18S	Forward: CAATCAAAACGTGTCCAACG
	Reverse: TAGCCAGGTTGCGAAGAACT

Immunostaining

The cells will have lost ABCG2 and PAX6 staining but will become positive for keratocan when stained with Kera C antibody.

Cryopreservation

This procedure is identical as that for limbal stem cells.

Variations and Applications of the Method

The corneal stromal stem cells have clonogenic and multipotent differentiation potential. These properties can be used to confirm the stem cell character of isolated cells.

Clonal growth

Stromal stem cells grow clonally in MJM. Trypsinized cells are counted and diluted in MJM to a concentration of 3 cells/mL. Plate 0.1 mL per well in half-area (A/2) 96-well microtitration plates. The ratio of 0.3 cells/well provides a very low chance that any well will have two cells. After 2 weeks, wells with colonies are marked and medium is changed. When confluent, the cells are trypsinized and expanded at 10^4 cells/cm^2. Cloning is recommended before differentiation.

Chondrogenic potential

Chondrocytes are never observed in mammalian eyes. The ability of cells to express cartilage-specific genes and gene products is therefore a strong marker for their multidifferentiation potential (and hence multipotent stem cell character). *Chondrocyte differentiation medium* (CDM) contains DMEM/MCDB-201, 2% FBS, 0.1 mM ascorbic acid-2-phosphate, 10^{-7} M dexamethasone, 10 ng/mL TGF-β 1, and 100 μg/mL sodium pyruvate. Cells (2 $\times$ 10^5) are resuspended in CDM and are pelleted in a 15-mL conical centrifuge tube. The medium is changed every 3 days. Pellets are cultured for 2 to 3 weeks. Messenger RNA for collagen II, aggrecan, and *cartilage oligomatrix protein* (COMP) can be detected by RT-PCR in stem cells cultured under the chondrogenic conditions but not in similar cultures of fibroblasts or keratocytes. Collagen II and COMP protein expression can also be detected with immunoblotting of pellet extracts.

Table. 8.2. Human primers for detection of cartilage-specific gene expression by RT-PCR

Collagen II	Forward: CCGGGCAGAGGGCAATAGCAGGTT
	Reverse: CAATGATGGGGAGGCGTGAG
COMP	Forward: ACAATGACGGAGTCCCTGAC
	Reverse: AAGCTGGAGCTGTCCTGGTA
Aggrecan	Forward: TGAGGAGGGCTGGAACAAGTACC
	Reverse: GGAGGTGGTAATTGCAGGGAACA

Neural differentiation

Stem cells are incubated under conditions that induce neural differentiation in Advanced D-MEM with 10 ng/mL *epithelial growth factor* (EGF), 10 ng/mL FGF-2 and 1 μM all-*trans* retinoic acid. Retinoic acid is added every 3 days, and the cells are kept 2 to 3 weeks to induce neurogenesis. RT-PCR from these cells will detect mRNA upregulation of both *glial fibrillary acidic protein* (GFAP) and *neurofilament protein*. Increases in GFAP and neurofilament proteins can also be detected by Western blotting. Immunofluorescent staining shows cells positive for neurofilament, GFAP, and β-tubulin III.

Table. 8.3. Human primers for detection of neural-specific gene expression by RT-PCR

Neurofilament protein	Forward: GAGGAACACCAAGTGGGAGA
	Reverse: CTCCTCCTCTTTGGCCTCTT
GFAP	Forward: ACTACATCGCCCTCCACATC
	Reverse: CAAAGGCACAGTTCCCAGAT

Culture of Human Corneal Endothelial Stem Cells

Scientists first attempted to isolate and culture *human corneal endothelial cells* (HCEC) about 40 years ago. At that time these cells were found to exhibit a very limited ability to divide in vitro. Contamination with stromal keratocytes caused technical problems with these studies. More recently, Joyce and colleagues cultured HCEC by stripping the Descemet membrane with HCEC intact to avoid contamination by stromal keratocytes. The presence of stem cells in human corneal endothelium remains controversial. Whikehart and co-workers (2005) suggested that endothelial stem cells may reside at the junctional region where the corneal endothelium meets the trabecular meshwork.

Telomerase activity and labeling by *bromodeoxyuridine* (BrdU) were detected in cells of this region after wounding. Yokoo et al. demonstrated that cells from human endothelium form spheres under conditions in which neural stem cells are known to form spheres, known as neurospheres. Mimura et al. (2005b,c) showed that presumptive stem cells isolated under these neurosphere-inducing conditions were effective for treatment of bullous keratopathy and corneal endothelium deficiency in rabbit models. These authors found that peripheral and central rabbit corneal endothelium contains a significant number of presumptive stem cells but the peripheral endothelium contains more and has a stronger self-renewal capacity than the central region as determined by sphere-forming assay.

The search for precursors/stem cells that can differentiate into corneal endothelium is of high clinical relevance. HCEC show little or no growth in vivo, and failure of endothelial function is a common cause of corneal opacity, thus generating the need for many corneal grafts.

An important tool in determining the ability of presumptive stem cells to differentiate into corneal endothelium is the presence of easily identifiable molecular markers of the HCEC cells. Foets et al. (1992a, b) showed that human corneal endothelial cells express *neural cell adhesion molecules* (N-CAM), *intercellular adhesion molecule*-1 (ICAM-1), and the transferrin receptor CD71 detected by monoclonal antibody OKT9. Cell border-associated *zonula occludens*-1 (ZO-1), a tight junction protein, has been extensively used to identify corneal endothelial morphology and integrity. Antibodies against some of these proteins, particularly N-CAM, anti-transferrin receptor monoclonal antibody OKT9, and tight junction protein ZO-1, are commercially available and thus provide the requisite tools to identify human corneal endothelial cells. The most important identification for differentiated corneal endothelial cells is function. The endothelium has a specific barrier and pump function for maintaining cornea transparency. The pump function of corneas as reconstructed in vitro with transplanted cells or endothelial sheet can be measured with an Ussing chamber. The most effective assay for pump function is the observation of corneal transparency in vivo in animals with cultured cells transplanted into the anterior chamber and allowed to settle on the Descemet membrane.

Isolation and Culture of Human Corneal Endothelial Cells

Protocol - 9. Isolation of human corneal endothelial cells

Reagents and materials

Sterile or aseptically prepared

1. Whole human cornea
2. CMF-Saline G
3. Dispase II, 2.4 U/mL: dilute to 1.2 U/mL in DMEM/F-12/GASP for use
4. Trypsin/EDTA
5. DMEM/F-12/GASP
6. DMEM/F-12/2FB: DMEM:Ham's F-12, 1:1, with 2% FBS
7. CMF/GASP
8. Scalpel or single-edged razor blades
9. Curved iris scissors, 11 cm (4-3/8 in.)
10. Jeweler's forceps, 10 cm (4 in.)
11. Corneal scissors, 19-mm blades, sharp tip
12. Colibri suturing forceps, 0.1 mm

Nonsterile

1. Variable-speed rocking mixer
2. Centrifuge, low speed, refrigerated
3. Tube roller apparatus

Procedure

1. Wash the cornea 3 times in PBSA, 5 min each time.
2. Trim off the residual sclera, conjunctiva, and iris.
3. Add 2 ml of 1.2 U/mL Dispase II, shaking gently at 4°C overnight.
4. Rotate the cornea in Dispase II at 4°C for 30 min.
5. Wash the cornea once in DMEM/F-12/GASP.
6. Under a dissection microscope, carefully remove trabecular meshwork, peel off the endothelium, and digest it in trypsin/EDTA at 37°C, 30 min.
7. Add the same volume of DMEM/F-12/2FB and disperse the cells.
8. Centrifuge at 400*g* for 10 min and discard the supernate.
9. Add 1 ml culture medium, gently resuspend the cells, and count in a hemocytometer.

Protocol - 10. Conventional culture of human corneal endothelial cells

Reagents and materials

1. Culture Medium: CECM
2. Trypsin/EDTA
3. Multiwell plate, 6-well

Procedure

1. Seed the cells into 1 well of 6-well plate precoated with FNC.
2. Change the medium every 3 days.
3. When the cells are 90% confluent, passage by trypsinization.

Protocol - 11. Culture of endothelial spheres

Reagents and materials

1. Culture medium: ESM
2. Trypsin/EDTA
3. Non-tissue culture-treated 24-well plate

Procedure

1. Trypsinize cells from Protocol 10.
2. Resuspend dissociated cells in culture medium at 10 cells/μL and plate on non-tissue culture coated 24-well plate.
3. After 7 days, cells are collected by centrifugation, dissociated by trypsin, and replated.

Characterization and Differentiation

The endothelial cells have very special morphology, forming a flat hexagonal cell sheet. H&E staining defines the cell shape. For ZO-1, N-CAM, OKT9 immunostaining, wash the dishes with PBSA once, fix with 100% acetone at –20°C for 10 min, wash with PBSA, then stain or store in 50% glycerol-PBSA at 4°C until staining.

Variations and Applications of the Method

Endothelial cells can also be cultured as explants. After peeling off the endothelium, cut it into small pieces of about 1-mm diameter and put the pieces into culture dishes with the endothelium side down. The cells will migrate out of the tissue. After confluence, passage the cells as described above.

Concluding Remark

Keratoplasty is currently the only effective method providing recovery of vision after corneal blindness. Although donated corneal tissue currently meets the needs of most recipients in the U.S., worldwide, 8 to 10 million individuals suffer from corneal blindness without access to therapy. Additionally, numerous individuals reject allogeneic corneal tissue, and the supply of donated corneas may soon be reduced by the increasing number of refractive surgeries, which render the corneas unsuitable for transplantation. Hence, there is significant interest in development of artificial and bioengineered corneas. Griffith et al. (1999) demonstrated that corneal equivalents generated from the three corneal cell layers mimic human corneas in key physical and physiological functions. These studies used immortalized cell lines transformed with retrovirus, making the engineered tissue unsuitable for transplantation. Focus has therefore turned to stem cells as a source of tissues for use in cell-based therapy and corneal tissue engineering. If we can use the stem cells from corneal epithelium, stroma, and endothelium to make artificial corneas suitable for clinical use, millions of patients suffering from corneal blindness will benefit.

9

Cartilage Stem Cells

The mature articular chondrocyte embedded in the cartilage matrix is a resting cell with no detectable mitotic activity and a very low synthetic activity. The markers of mature articular chondrocytes are type II collagen (COL2A1), other cartilage-specific collagens IX (COL9) and XI (COL11), the large aggregating proteoglycan aggrecan, and link protein. Chondrocytes also synthesize a number of small proteoglycans such as biglycan and decorin and other specific and nonspecific matrix proteins both in vivo and in vitro. As the single cellular constituent of adult articular cartilage, chondrocytes are responsible for maintaining the cartilage matrix in a low turnover state of equilibrium. In mature cartilage, the chondrocytes synthesize matrix components very slowly. The turn-over of collagen in normal adult articular cartilage has been estimated to occur with a half-life of longer than 100 yr and the half-life of aggrecan subfractions is in the range of 3–24 yr, whereas the glycosaminoglycan constituents on the aggrecan core protein are more readily replaced. Furthermore, normal chondrocyte metabolism *in situ* occurs in low oxygen tension and is remote from a vascular supply. Thus, it is not surprising that changes in expression of these cartilage matrix constituents occur when the chondrocytes are isolated and placed in monolayer culture, where they increase synthetic activity by several orders of magnitude.

Primary cultures of articular chondrocytes isolated from various animal and human sources have served as useful models for studying the mechanisms controlling responses to growth factors and cytokines. Early attempts to culture chondrocytes were frustrated by the tendency of these cells to "*dedifferentiate*" in monolayer culture and their inability to proliferate in suspension culture where cartilage-specific

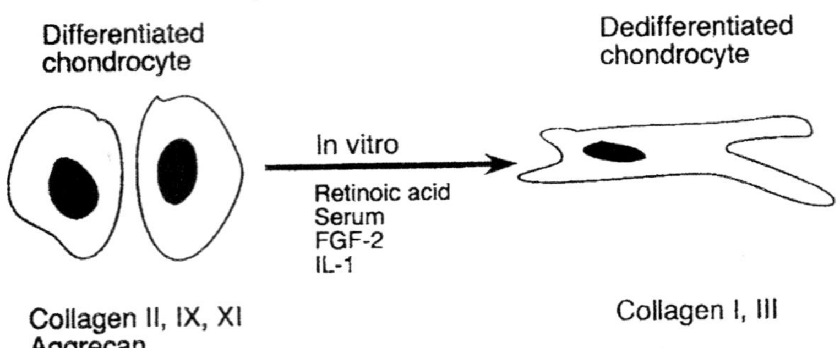

Fig. 9.1. Schematic representation of the "switch" from the differentiated to the dedifferentiated chondrocyte phenotype that occurs during culture and in response to certain cytokines and growth factors in vitro.

phenotype was maintained. In monolayer culture, chondrocytes maintain a rounded, polygonal morphology, but there may be a progressive change to a fibroblast-like morphology with passage of time, especially after subculture. High-density monolayer cultures maintain the cartilage-specific phenotype until they are subcultured, although gene expression of type II collagen is generally more labile than that of aggrecan. Concomitant with the loss of differentiated phenotype, which can be accelerated by plating the cells at low densities or by treatment with cytokines such as interleukin 1 (IL-1)or retinoic acid, the chondrocytes acquire some, but not all, characteristics of the fibroblast phenotype, i.e., type I collagen.

Because the stability of the phenotype of isolated chondrocytes is critically dependent on *cell shape* and *cell density*, high-density micromass cultures are useful if sufficient numbers of chondrocytes are isolated, particularly for studying proteoglycan biosynthesis. It is also possible to expand the cultures through a limited number of subcultures and "*redifferentiate*" the cells in fluid or gel suspension culture systems, where the chondrocytes regain morphology and the cessation of proliferation is associated with increased expression of cartilage-specific matrix proteins. Culture systems that support chondrocyte phenotype include suspension culture in spinner flasks, in dishes coated with a nonadherent substrates, in pellets, and in three-dimensional matrices such as collagen gels, agarose, alginate, or collagen sponges. Serum-free defined media of varying compositions, but usually including insulin, have also been used, frequently in combination with the other culture systems mentioned above.

The use of chondrocytes of human origin has been problematical, because the source of the cartilage cannot be controlled, sufficient

Table 9.1. Proteins synthesized by mature chondrocytes

Collagens
Type II
Type IX
Type XI
Type VI
Types XII, XIV
Proteoglycans
Aggrecan
Versican
Link protein
Biglycan (DS-PGI)
Decorin (DS-PGII)
Epiphycan (DS-PGIII)
Fibromodulin
Lumican
PRELP (proline/arginine-rich and leucine-rich repeat protein)
Chondroadherin
Perlecan
Lubricin (SZP)
Other noncollagenous proteins (structural)
Cartilage oligomeric matrix protein (COMP; thrombospondin-5))
Thrombospondin-1 and -3
Cartilage matrix protein (matrilin-1); matrilin-3
Fibronectin
Tenascin-c
Cartilage intermediate layer protein (CILP)
Fibrillin
Elastin
Other noncollagenous proteins (regulatory)
S-100
Chondromodulin-I (SCGP) and -II
Glycoprotein (gp)-39, YKL-40
Matrix Gla protein (MGP)
CD-RAP (cartilage-derived retinoic acid-sensitive protein)
Growth factors
Membrane-associate proteins
Integrins ($\alpha1\beta1$, $\alpha2\beta1$, $\alpha3\beta1$, $\alpha5\beta1$, $\alpha6\beta1$, $\alpha10\beta1$, $\alpha v\beta3$, $\alpha v\beta5$)
Anchorin CII (annexin V)
CD44
Syndecan-3

numbers of cells are not readily obtained from random operative procedures, and the phenotypic stability and proliferative capacity of adult human chondrocytes are lost more quickly upon expansion in serial monolayer cultures than in cells of juvenile human or embryonic or postnatal animal origin. Alternatively, explant cultures of human, but usually bovine, articular cartilage where the chondrocytes remain encased within their own extracellular matrix, have been used as in vitro models to study cartilage biochemistry and metabolism. However, many experimental manipulations are done more easily using isolated chondrocytes. Recent studies have focused on adult human articular chondrocytes as target cells for immortalization, using immortalizing antigens such as SV40-TAg, human papilloma virus type 16 (HPV-16) early function genes E6 and E7 and telomerase. Strategies that maintain high cell density and decrease cell proliferation must also be applied to immortalized chondrocyte cell lines, since stable integration of immortalizing genes disrupts normal cell cycle control but does not stabilize expression of the type II collagen gene.

This chapter will focus on strategies for isolation, culture, and characterization of isolated human articular chondrocytes and will also describe approaches for using different chondrocyte culture systems for evaluating chondrocyte phenotype and studying the regulation of chondrocyte functions.

Materials

Isolation and Culture of Human Chondrocytes

1. Growth medium for chondrocytes: Mix equal volumes of Dulbecco's modified Eagle's medium (DMEM), high glucose (4.5 g/L), and Ham's F-12, 11 mixture, both with L-glutamine and without HEPES, to give final concentrations of 3.151 g/L glucose, 365 mg/L L-glutamine/7.36 mg/L L-glutamic acid, and 110 mg/L sodium pyruvate. Add 10% FCS immediately before use.
2. Dulbecco's phosphate-buffered Ca^{2+}- and Mg^{2+}-free saline *phosphate-buffered serum* (PBS).
3. Trypsin-ethylenediaminetetraacetic acid (EDTA) solution: 0.05% trypsin and 0.02% EDTA in *Hanks' balanced salt solution* (HBSS) without Ca^{2+} and Mg^{2+}.
4. Hank's balanced salt solution (HBSS) with Ca^{2+} and Mg^{2+}.
5. Serum substitutes for experimental incubations: Nutridoma-SP is provided as sterile concentrate (100X, pH 7.4; storage at 15–25°C protected from light). Dilute 1100 (v/v) with sterile DMEM/Ham's

F-12, without FCS, and use immediately. ITS+ is an alternative serum substitute.

Except where specified, cell culture reagents can be obtained from a number of different suppliers, including Life Technologies, Gaithersburg, MD; Cambrex, Inc.,Walkersville, MD; Sigma, St. Louis, MO; or Intergen, Purchase, NY. Reserve serum testing, which is offered by these suppliers, is recommended for selection of lots of FCS that maintain chondrocyte phenotype. FCS and trypsin-EDTA are stored at –20°C, but should not be refrozen after thawing for use.

6. Enzymes for cartilage digestion:
   (a) Hyaluronidase (Sigma): 1 mg/mL in PBS: Prepare freshly and filter through a sterile 0.22-μm filter.
   (b) Trypsin: 0.25% in HBSS without Ca^{2+} and Mg^{2+}.
   (c) Collagenase (bacterial, clostridiopeptidase A) or Collagenase D: 3 mg/mL in DMEM with 10% FCS for articular cartilage or serum-free for costal cartilage: Prepare freshly in ice-cold DMEM and filter immediately through a sterile 0.22-μm filter.

Suspension Culture Systems for Chondrocytes

1. Agarose-coated dishes: Weigh out 10 mL high-melting-point agarose in an autoclavable bottle and add 100 mL of dH_2O. Autoclave with cap tightened loosely, allow to cool to ~approx 55°C, and pipet quickly into culture dishes (1 mL/3.5-cm well of 6-well plate, 3 mL/6-cm dish, or 9 mL/10-cm dish). Allow the gel to set at 4°C for 30 min and wash the surface two or three times with PBS. Plates may be used immediately or wrapped tightly with plastic or foil to prevent evaporation and stored at 4°C.
2. Poly-HEMA (poly-2-hydroxyethyl-methacrylate)-coated dishes: Prepare a 10% (w/v) solution by dissolving 5 g of poly-HEMA in 50 mL of ethanol in a sterile capped bottle or centrifuge tube. Leave overnight at 37°C with gentle shaking to dissolve polymer completely. Centrifuge the viscous solution for 30 min at 2000*g* to remove undissolved particles. Layer the polyHEMA solution on dishes at 0.3 mL/well of 6-well plate or 0.9 mL/6-cm dish and leave with lids in place to dry overnight in a tissue-culture hood. Expose open dishes to bactericidal ultraviolet light for 30 min to sterilize.
3. Agarose: Autoclave 2% (w/v) low-gelling-temperature agarose in dH_2O, cool to 37°C, and dilute with an equal volume of 2X

DMEM containing 20% FCS either without cells or with a chondrocyte suspension.

4. Alginate (Keltone LVCR, NF). *Low viscosity* (LV) alginate is used generally. Request LVCR for more highly purified preparation.
   (a) Prepare 1.2% (w/v) solution of alginate in 0.15 *M* NaCl.
   (b) Dissolve alginate in a 0.15 *M* NaCl solution, heating the solution in a microwave oven until it just begins to boil. Swirl and heat again two or three times until the alginate is dissolved completely. Allow the solution to cool to about 37°C and sterile filter. Filtering when warm permits the viscous solution to pass through the filter.
   (c) Prepare 102 m*M* $CaCl_2$ and 0.15 *M* NaCl solutions in tissue culture bottles and autoclave.
   (d) Prepare 55 m*M* Na citrate, 0.15 *M* NaCl, pH 6.0, sterile filter, and store at 4°C. Make fresh weekly.
5. Three-dimensional (3D) scaffolds: Several types of scaffolds are available commercially, including the following:
   (a) Gelfoam, sterile absorbable collagen sponge, purchased as sponge-size 12–7 mm (2 cm × 6 cm × 7 mm).
   (b) BD Three Dimensional Collagen Composite Scaffold: Contains a mixture of bovine type 1 and type III collagens and is provided as 3D scaffolds with 48-well plates.
   (c) BD Three Dimensional OPLA Scaffold: Contains a synthetic polymer synthesized from D,D-L,L polylactic acid and is provided as 3D scaffolds with 48-well plates.
6. Cell lysis solution for recovery of cells from scaffolds: 0.2% v/v Triton X-100, 10 m*M* Tris-HCl, pH 7.0, 1 m*M* EDTA.
7. Collagenase solution for recovery of cells from scaffolds: 0.03% (w/v) collagenase in HBSS.

Analysis of Matrix Protein Synthesis

Alcian blue staining

1. 2.5% glutaraldehyde (diluted from 50% solution; Sigma) in 0.4 *M* $MgCl_2$ and 25 m*M* sodium acetate, pH 5.6.
2. Alcian blue 8GX (Sigma). Dissolve in the 2.5% glutaraldehyde solution to give final concentration of 0.05%. Filter the solution through Whatmann paper (or coffee filter).
3. Washes: 3% acetic acid solutions without and with 25 and 50% ethanol.

Collagen typing

1. L-[5-^{3}H]proline (1 mCi/mL; specific activity >20 Ci/mmol) at 25 μCi/mL in serum-free culture medium supplemented with 50 μg/mL ascorbate and 50 μg/mL β-aminoproprionitrile fumarate (β-APN). Sterile filter 10X solution of ascorbic acid and β-APN (5 mg of each dissolved in 10 mL serum-free culture medium), dilute in medium at 1/10 (v/v) to give the volume required for the incubation, and add 25 μL of [^{3}H]proline per mL using a sterile pipet tip.
2. Pepsin/acetic acid solution: Dissolve 2 mg of pepsin in 1 mL dH_2O, then add 58 μL glacial acetic acid per each milliliter of solution and cool on ice.
3. Gel sample buffer (GSB): 0.1 M Tris-HCl, pH 7.6, 3% (w/v) SDS, and 16%(v/v) glycerol. Reagents for sodium dodecyl sulfate-polyacrylamide gel electrophoresis (SDS-PAGE) and autoradiography.
4. Loading dye: 1% (w/v) bromophenol blue (BPB; sodium salt).
5. Tris-glycine/SDS-5% gradient polyacrylamide or 7–15% gradient gels and Laemmli buffer system.

Proteoglycan synthesis

1. Biosynthetic labeling of proteoglycans: [^{35}S]sodium sulfate (2 mCi/mL; specific activity >1000 Ci/mmol). Add to culture medium at 50 μCi/mL.
2. 7 *M* urea.
3. DE52 columns, 3.5 . 12-cm containing 4 mL of DEAE-cellulose.
4. Guanidine extraction buffer: 4.0 *M* guanidine-HCl, buffered with 50 m*M* sodium acetate, and containing 10 m*M* disodium EDTA. Add immediately before use 100 m*M* 6-aminocaproic acid, 2.5 m*M* benzamidine HCl, 5 m*M* N-ethylmaleimide, and 0.25 m*M* phenylsulfonyl fluoride (PMSF) from 100X stock solutions in absolute ethanol. The 2X guanidine extraction buffer is prepared at twice the concentrations above.
5. SDS-PAGE: Pre-cast Tris-glycine SDS-polyacrylamide 4–20% gradient gels. Use the GSB and BPB solutions for loading the samples and the Lemmli buffer system.

Immunocytochemistry

1. Fixative: 2% paraformaldehyde in 0.1 *M* cacodylate buffer, pH 7.4. Dissolve 10 g paraformaldehyde in 150 mL dH_2O in Ehrlenmeyer flask on hot plate in fume hood (do not exceed 65°C).

Add ~approx 2 mL of 1 *N* NaOH while stirring, and stir until solution is clear. Let solution cool for 15 min. Add 250 mL 0.2 *M* cacodylate buffer, pH 7.4, and adjust pH if necessary.

Western blotting

1. RIPA buffer: 50 m*M* Tris-HCl, pH 7.4, 150 m*M* NaCl, 0.5% sodium deoxycholate, 0.1% SDS, and 1% NP-40. At time of use, add Complete, EDTA-free proteinase inhibitor cocktail.
2. Tris-glycine/SDS-polyacrylamide gels at polyacrylamide concentration appropriate for the size of matrix protein to be analyzed.
3. Nylon-supported nitrocellulose membranes, Immobilon-P, 0.45-μm.
4. Transfer buffer: 25 m*M* Tris, pH 7.6, 192 m*M* glycine, 20 % (v/v) methanol.
5. Tris-buffered saline (TBS)/Tween (TBST): 20 m*M* Tris-HCl, 137 m*M* NaCl, 0.1% (v/v) Tween-20, pH 7.6. Add 5% (w/v) nonfat dry milk (Carnation) as required.
6. Primary antibody and horseradish peroxidase-conjugated secondary antibody: Dilute according to the supplier's instructions in TBST containing 5% (w/v) nonfat dry milk.
7. Enhanced Chemiluminescence ECL Western blotting analysis system.

Antibodies

Antibodies that detect human collagens and proteoglycans that are specific to cartilage are available commercially from Southern Biotechnology Associates, Inc., IBEX Technologies, Inc., and Chemicon International. Some antibody preparations are useful for developing quantitative *enzyme-linked immunosorbent assay* (ELISA) assays.

Analysis of mRNA

1. RNA extraction kit: The TRIzol reagent or RNeasy Mini Kit are suitable for extraction of total RNA from chondrocytes.
2. Sterile, RNase-free solutions, polypropylene tubes, and other materials.

Analysis of Gene Expression by Transfection of Regulatory DNA Sequences

1. EndoFree Plasmid Maxi Kit.
2. Lipid-based transfection reagent such as LipofectAMINE PLUS Reagent or FuGENE 6.
3. Serum-free culture medium for transfections: Opti-MEM or DMEM/F-12 (test for optimal transfection efficiency).

4. Passive Lysis Buffer.
5. Coomassie Plus Protein Assay Reagent.
6. Luciferase Assay System.
7. Dual-Luciferase Reporter Assay with the pRL-TK Renilla luciferase control vector.
8. Adenovirus producer cell line: 293 (ATTC CRL 1573; transformed primary human embryonic kidney).

METHODS

The methods are described as follows: (i) the isolation of human chondrocytes from cartilage and their primary culture in monolayer and passaging, (ii) suspension culture systems for maintaining chondrocyte phenotype, analysis of the (iii) synthesis and (iv) mRNA expression of cartilage-specific matrix proteins, and (v) analysis of gene expression by transfection of regulatory DNA sequences.

Isolation and Culture of Human Chondrocytes in Monolayer

1. Human adult articular cartilage is obtained, after Institutional Review Board approval, from the knee joints or hips after orthopaedic surgery for joint replacement or reconstruction, or at autopsy, and dissected free from underlying bone and any adherent connective tissue.
2. Place slices of cartilage in a 10-cm dish and wash several times with PBS. Incubate slices at 37°C in hyaluronidase for 10 min followed by 0.25% trypsin for 30–45 min with two or three washes in PBS after each enzyme treatment. Use approx 10 mL of proteinase solution for digestion of each gram of tissue.
3. Add collagenase solution, chop the cartilage in small pieces using a scalpel blade, and incubate at 37°C overnight (18–24 h) for articular cartilage and up to 48 h for costal cartilage until the cartilage matrix is completely digested and the cells are free in suspension. Break up any clumps of cells by repeated aspiration of the suspension through a 10-mL pipet or a 12-cc syringe without a needle.
4. Transfer cell suspension to a sterile 50-mL conical polypropylene tube and wash the plate with PBS to recover remaining cells and combine in tube. Centrifuge cells at 1000*g* in a bench-top centrifuge for 10 min at room temperature and wash the cell pellet three times with PBS, resuspending cells each time and centrifuging.
5. Resuspend the final pellet in DMEM/F12 containing 10% FCS, perform cell count with a Coulter counter or hemacytometer, and

bring up to volume with culture medium to give 1 × 10^6 cells per mL. For monolayer culture, plate cells at 2.5 × 10^4/cm^2 in dishes or wells containing culture medium, and agitate without swirling to distribute the cells evenly. Incubate at 37°C in an atmosphere of 5% CO_2 in air with medium changes after 2 d and every 3 or 4 d. Primary of adult articular chondrocytes incubated in the absence or presence of IL-1β.

Table 9.2. Culture vessel area vs chondrocyte number required for plating density of approx 2.5 × 10^4 cells/cm^2

Diameter	*Area (cm^2)*	*No. of cells plated*
16-mm well (24-well)	2	50,000
2.2-cm well (12-well)	3.8	100,000
3.5-cm well (6-well)	10	250,000
6-cm plate	28	750,000
10-cm plate	79	2 × 10^6

6. Preparation of subcultured cells: Remove culture medium by aspiration with a sterile Pasteur pipet attached to a vacuum flask and wash with PBS. Add trypsin-EDTA (1 mL/10-cm dish) and incubate at room temperature for 10 min with periodic gentle shaking of dish and observation through microscope to ensure that cells have come off the plate. If significant numbers of cells remain attached, continue the incubation for a longer time (≤ 20 min) or at a higher temperature (37°C) and/or scrape the cell layer with a sterile plastic scraper or syringe plunger. Repeatedly aspirate and expel the cell suspension into the plate using a 5- or 10-mL pipet containing culture medium, and then transfer to a sterile conical 15- or 50-mL polypropylene tube. Perform cell counts or determine the split ratio required (usually one dish into two, or 1:2, for adult articular chondrocytes, or 1:5 for more rapidly growing, denser juvenile chondrocytes). Distribute equal volumes of the cell suspension in dishes or wells that already contain culture medium, rocking plates back-and-forth (not swirling) immediately after each addition to ensure uniform plating density.
7. For experiments, plate cells in dishes or wells at a concentration of 2 × 10^6 cells/cm^2 in DMEM/F12 containing 10% FCS. Remove the growth medium, when the cultures are confluent. Add serum-free medium containing a serum substitute, such as Nutridoma-SP, followed 18–24 h later by the test agent of interest. Continue

incubation at 37°C for short-term time courses of 0.25–24 h or longer time courses of several days.

Suspension Culture Systems for Chondrocytes

Chondrocytes in monolayer culture are susceptible to loss of phenotype during prolonged culture and particularly after subculture. Thus, it is necessary to use culture conditions that maintain differentiated chondrocyte features or that permit redifferentiation. Freshly isolated chondrocytes may be cultured immediately in suspension, where they do not proliferate, or they may be expanded in monolayer culture and placed in suspension culture after several passages.

Fluid suspension cultures on agarose- or polyHEMA-coated dishes

1. Trypsinize monolayer cultures, spin down cells, wash with PBS, centrifuge, and resuspend in culture medium containing 10% FCS at 1×10^6 cells/mL.
2. Transfer chondrocyte suspension to dishes that have been coated with 1% agarose or with 0.9% polyHEMA and culture for 2–4 wk. The cells first form large clumps that begin to break up after 7–10 d and eventually form single-cell suspensions.
3. Change the medium weekly by carefully removing the medium above settled cells while tilting the dish, centrifuging the remaining suspended cells, and replacing them in the dish after resuspension in fresh culture medium.
4. To recover cells for direct experimental analysis, for redistribution in agarose- or polyHEMA-coated wells, or for culture in monolayer, transfer the cell suspension to 15- or 50-mL conical tubes, gently washing the agarose surface at least twice with culture medium to recover remaining cells, and spin down and resuspend cells in an appropriate volume of culture medium for plating or in extraction buffer for subsequent experimental analysis.

Suspension culture within agarose

1. Precoat plastic tissue culture dishes with cell-free 1% agarose in culture medium (0.5 mL/3.5-cm, 1.5 mL/ 6-cm, or 4.5 mL/10-cm dish) and allow to gel at room temperature. Add the same volume of 1% agarose in medium containing chondrocytes at a density of 1–4 $\times 10^6$ cells/mL of gel, incubate at 37°C for 20–30 min to allow the cells to settle, and leave at room temperature until the agarose forms a gel. Add culture medium containing 10% FCS and incubate at 37°C with medium changes every 3–4 d.

2. After incubations with test reagents and/or radioisotopes in minimal volumes of appropriate culture medium, the whole cultures may be stored frozen or medium and gel treated separately. For subsequent analysis, add appropriate guanidine extraction buffer directly to the gel (for proteoglycan or RNA extraction) or whole cultures may be adjusted to 0.5 *M* acetic acid, treated with pepsin, and neutralized, as described below for analysis of collagens. To remove agarose and debris, the samples are centrifuged in a high-speed centrifuge at >10,000*g* at 4°C.

Alginate bead cultures

1. Trypsinize several 10-cm plates and wash the cells with PBS. Determine the cell count with a hemacytometer and pellet the cells.
2. Resuspend the pellet in a 1.2% solution of alginate in 0.15 *M* NaCl at a concentration of 1–4 × 10^6 cells per mL. Slowly express the alginate suspension in a dropwise manner through a 10-cc syringe equipped with a 22-gauge needle into a 50-mL polypropylene centrifuge tube containing 40 mL of 102 m*M* $CaCl_2$. Allow the beads to polymerize in the $CaCl_2$ solution for 10 min and wash twice with 25 mL of 0.15 *M* NaCl. The alginate beads should not be washed in PBS, as they will become cloudy.
3. Resuspend the beads at 7–15 beads per mL in growth medium supplemented with 25 μg/mL Na ascorbate and decant to a culture dish or flask. Culture in DMEM/F12 with medium changes every 3 d, carefully pipetting the spent culture medium from the top of the settled beads.
4. At the end of the culture period (one to several weeks), add the appropriate guanidine extraction buffer or centrifuge at 500*g* for 10 min to recover the chondrocytes with pericellular matrix.
5. Alternatively, to recover cells from alginate, carefully aspirate the medium from the cultures and wash twice with PBS. Depolymerize the alginate by adding three volumes of a solution of 55 m*M* Na citrate/0.15 *M* NaCl and incubate at 37°C for 10 min. Aspirate the solution over the surface of the dish several times to dislodge adherent cells (the cells are sticky) and transfer the suspension to a 50-mL centrifuge tube. Because the solution is quite viscous, centrifuge the cells at 2000*g* for a minimum of 10 min to completely pellet the cells. Wash the cells twice with PBS before using them for further analysis.

Culture on 3-D scaffolds

1. Gelfoam: Use sterile scalpel blade to cut into pieces of 1 ×1 × 0.5 cm³ and place in wells of sterile six-well plates. Inoculate by dropping 50 μL of growth medium containing 10^6 cells on each sponge. Place in incubator for 1.5–2 h, then add 100 μL medium and culture for an additional 1–3 h. Add medium to cover and continue incubation overnight or longer.
2. BD 3D Collagen Composite or OPLA Scaffolds: Place scaffolds (0.5 cm³) in the 48-well plates provided, in 96-well plates, or other plate as required. Seed scaffolds by dropping 100 μL of growth medium containing 1–5 × 10^4 cells. Incubate for 1 h, add 150 μL of medium to each scaffold, and incubate for 1.5–3 h. Add medium as required for further culture and experimental conditions.
3. Recovery of cells from scaffolds for analysis:
   (a) Prepare cell lysates for DNA analysis using 250–500 μL of cell lysis solution [0.2% (v/v) Triton X-100, 10 m*M* Tris-HCl, pH 7.0, 1 m*M* EDTA] per scaffold in 1.5-mL tube. Freeze samples at –70°C and subject to two freeze/thaw cycles, thawing at room temperature for 45–60 min. Break up scaffolds with pipet tip, centrifuge, and transfer lysates to fresh tubes. The cell lysates may be analyzed using the Picogreen Assay Kit according to the manufacturer's protocol.
   (b) Recover cells for RNA extraction and other analyses by treatment of minced scaffolds with 0.03% (w/v) collagenase in HBSS for 10–15 min at 37°C. Collect cells by centrifugation, wash with PBS, and add appropriate extraction buffer to the final pellet.

Analysis of Matrix Protein Synthesis

Chondrocyte culture models are used to examine the effects of cytokines and growth/differentiation factors on the synthesis of chondrocyte phenotypic markers by staining the glycosaminoglycans with alcian blue, characterizing the collagens and proteoglycans synthesized, and perform immunohistochemistry using specific antibodies against these proteins.

Alcian blue staining

1. Monolayer cultures: Remove culture medium from confluent cultures that have been incubated in the presence of 50 mg/mL ascorbic acid for at least 4 h and wash with PBS. Add alcian blue/glutaraldehyde solution at room temperature for several hours,

remove excess stain by washing with 3% acetic acid, and store cultures in 70% ethanol for subsequent examination by light photomicrography.

2. Alginate bead cultures:
   (a) Using a 25-mL pipet, transfer five beads to a 12 × 15 cm tube, and wash twice with 2 mL PBS. Add 1 mL alcian blue stain, and 50 μL of a 50% glutaraldehyde solution. Store for 24 h at 4°C.
   (b) Aspirate the stain from the beads, and wash twice with 2 mL of 3% (v/v) acetic acid. Destain the beads with rocking at room temp for 5 min sequentially with 2 mL of each of the following solutions: (i) 3% acetic acid, (ii) 3% acetic acid/25% ethanol, and (iii) 3% acetic acid/50% ethanol. Store the beads in 70% ethanol.
   (c) Before photographing the stained beads, gently flatten beneath a glass cover slip, taking care not to disrupt the alginate matrix. Alternatively, the beads may be embedded and sectioned prior to photography.

Collagen typing

1. Biosynthetic labeling of collagens: Remove serum-containing culture medium, wash with serum-free medium, and add [^{3}H]proline at 25 μCi/mL for a further 24 h in serum-free culture medium supplemented with 50 μg/mL ascorbate and 50 μg/mL β-APN (or without β-APN to retain collagen in the pericellular matrix). Remove culture medium and store at –20°C. Wash cell layer with PBS and solubilize by adding equal volumes of serum-free culture medium and 1 *M* ammonium hydroxide (an aliquot may be analyzed for DNA).
2. Collagen typing: To analyze pepsin-resistant collagens, add pepsin/acetic acid solution to an equal volume of either labeled culture medium or solubilized cell solution for 16 h at 4°C, lyophilize, redissolve in 2X SDS sample buffer, and neutralize with 1 μL additions of 2 *M* NaOH to titrate the color from yellow-green to blue (but not to violet). To analyze procollagens and fibronectin, add 2X SDS sample buffer containing 0.2% β-ME to an equal volume of the culture medium. Heat samples to boiling for 10 min and load on SDS gels (5% acrylamide running gels or 7–15% gradient gels) that include a radiolabeled rat tail tendon collagen standard in one lane. Perform delayed reduction with 0.1% β-ME on pepsinized samples to distinguish α1(III) from α1(I or II)

collagens. Absence of the α2(I) collagen band generally indicates the absence of type I collagen synthesis. In cultures containing a mixture of type I and type II ollagen, definitive identification of these collagens requires Western blotting using specific antibodies.

Proteoglycan synthesis

1. Monolayer cultures: Aspirate the culture medium and wash the cell layer with PBS. Add DMEM/F12 containing 10% FCS supplemented with [^{35}S]sulfate at 50 μCi/mL and 25 μg/mL ascorbate. Incubate at 37°C for 18 h. Remove the conditioned medium to a 15-mL polypropylene tube and wash the cell layer three times with PBS.
   (a) Medium extraction and purification: Add an equal volume of 7 *M* urea to the medium. Mix and count 10 μL in a liquid scintillation counter. Pass up to 4 mL through DEAE-Cellulose (DE52) column, which has been preequilibrated with 7 *M* urea, to remove unincorporated cpm. Elute proteoglycans (PGs) with 2 mL of 4 *M* guanidine extraction buffer. To 500 μL of column eluate, add two volumes of 100% ethanol and precipitate PGs for 2 h at –20°C. Spin at 10,000*g* for 20 min at 4°C. Wash final pellet with 70% ethanol.
   (b) Cell layer extraction and purification: Add 4 *M* guanidine extraction buffer to cell layer e.g., 2 mL/25-cm^2 flask) and extract at 4°C for 24 h with rocking. Transfer extract to a 2-mL screw cap microcentrifuge tube and spin at 10,000*g* for 20 min at 4°C to pellet particulate material from the sample. Remove supernatant to a fresh 2-mL tube and count 10 μL in a liquid scintillation counter. (Also, 10–50 μL aliquots may be taken at this point for DNA analysis.) Take 250 μL for precipitation of PGs by addition of three volumes of 100% ethanol at –20°C for 2 h (or overnight for alginate extracts), spin at 10,000*g* for 20 min at 4°C, and wash final pellet with 70% ethanol. Dry pellet at room temperature for 30 min.
2. Alginate cultures: Aspirate the medium from a culture containing 50 beads in a 25-cm^2 flask cultured on end. Add 4 mL of growth medium supplemented with [^{35}S]sulfate at 50 μCi/mL and 25 μg/mL ascorbate. Incubate at 37°C for 18 h. Remove the conditioned medium to a 12-mL polypropylene tube and wash the beads three times with 5 mL of PBS (5 min/wash).
   (a) Medium extraction and purification: Perform as described for monolayer cultures.

(b) Extraction from alginate beads and purification: Transfer the beads to a 15-mL polypropylene tube, extract radiolabeled PGs with 2 mL of 4 *M* guanidine extraction buffer at 4°C for 24 h with rocking. Transfer extract to a 2-mL screw cap microcentrifuge tube and proceed as described for monolayers.

3. Characterization of proteoglycans in monolayer and alginate cultures:
   (a) Sulfate incorporation into proteoglycans: Dissolve ethanol-precipitated pellet in 100 μL of 1X GSB. Count 3 μL in a liquid scintillation counter and calculate the cpm incorporated, after accounting for dilutions, against the concentration of DNA in the cell extract. The [^{35}S]sulfate incorporation may also be determined, after passing the guanidine extracts over Sephadex G-25M in PD 10 columns and eluting under dissociative conditions, by scintillation counting.
   (b) SDS-PAGE analysis: Take a volume of cell or medium extract in GSB that corresponds to 0.25 μg of DNA in the cell extract and add DTT to a final concentration of 0.5 m*M* and 1% bromophenol blue to a final concentration of 0.1%. Heat 5 min at 100°C and store remaining sample at -20°C. Electrophorese ~approx 20,000 cpm on a 4–20% polyacrylamide gradient gel. Fix the gel in acetic acid/methanol for 1 h, dry, and expose to film at -80°C. Visualize radiolabeled PGs by autoradiography, as shown above.

Immunocytochemistry

1. Plate cells in plastic Lab-Tech four-chamber slides at 6 × 10^4 cells/chamber in culture medium containing 10% FCS. Add 25 μg/mL ascorbic acid with the first medium change and daily thereafter.
2. When the cultures have reached confluence, add the desired test reagents. At the end of the incubation period, carefully wash the chambers three times with PBS, and fix the cells with 2% paraformaldehyde in 0.1 *M* cacodylate buffer, pH 7.4, for 2 h at 4°C.
3. Rinse twice with 0.1 *M* cacodylate buffer. Add antibodies that recognize human type II collagen, aggrecan, and so on to different chambers at concentrations recommended by the supplier. Incubate separate chamber slides with chondroitinase ABC for 30 min at 37°C prior to addition of monoclonal antibodies in order to expose epitopes.
4. Visualize the staining by incubation with a gold-conjugated secondary antibody followed by silver enhancement.

Western blotting

1. Plate cells in six-well tissue culture plates at a density of 0.25×10^6 cells per well in culture medium containing 10% FCS. Add 25 μg/mL ascorbic acid with the first medium change and daily thereafter. When the cultures have reached confluence, add the desired test reagents.
2. At the end of the incubation, remove medium, wash cell layers with 3 mL/well of PBS, and add 0.2 mL of ice-cold RIPA buffer containing freshly added proteinase inhbitors. Extract for 15 min at 4°C with gentle rocking, scrape and transfer cell lysate to a microcentrifuge tube. Incubate 30–60 min on ice, microcentrifuge at 10,000*g* for 10 min at 4°C, and transfer supernatant to new tube, discarding pellet.
3. Mix cell lysate (20–40 μg of protein) with equal volume of 2X GSB, boil for 2–3 min, and electrophorese on a Tris-glycine/SDS polyacrylamide pre-cast gel, and transfer to Immobilon-P membrane by electroblotting with a transfer current of 100 V for 1 h.
4. Block membrane with 10 mL TBST containing 5% nonfat powdered milk at 4°C for 1 h with rocking, and wash three times (10 min/wash) in 10 mL TBST.
5. Incubate membrane with the antibody at the appropriate dilution in 5 mL of TBST containing 5% nonfat powdered milk for 1 h at room temperature with rocking. Wash membrane three times (5 min/wash) with TBS, 0.05% Tween-20. Incubate with secondary antibody diluted in TBST for 45 min and repeat washes as above, followed by one wash with TBS for 5 min.
6. Detect bound antibody by enhanced chemiluminescence according to the manufacturer's protocol and expose to film for autoradiography.

Extraction and Analysis of RNA

1. Recovery of cells for extraction: Trypsinize monolayer cultures as described above, and add medium containing 10% FCS to the suspension to inactivate the trypsin. Depolymerize the alginate cultures as described above. Transfer cell suspension to sterile, RNase-free polypropylene tube of appropriate size, centrifuge at 1000*g* at 4°C, washing two or three times with ice-cold PBS.
2. Extract total RNA using TRIzol reagent, RNeasy Mini Kit, or other method as preferred. Add extraction buffer to final pellet, vortex vigorously, and transfer to RNase-free 1.2-mL centrifuge tube. Continue according to the manufacturer's instructions. The

final pellets are usually washed with 75% ethanol and centrifuged in a speed-vac apparatus, but not to dryness. Dissolve pellets in nuclease-free water.

3. Read ODs at 240, 260, and 280. Use OD 260 to calculate RNA concentration. The final preparations should give yields of approx 10 μg of RNA per 1 × 10^6 cells with the appropriate A_{260}:A_{280} ratio of approx 2.0. Store at –20°C in nonself-defrosting freezer or at –80°C.
4. Analyze mRNAs by Northern blotting, semiquantitative RT-PCR, or real-time PCR by published methods. The Northern blots indicate that type II collagen gene expression is relatively stable in primary cultures of human chondrocytes, at least through the first 12 d of culture. Type I collagen gene expression is low, but TGF-β is a potent stimulant. This experiment also illustrates that the effects of TGF-β on type II collagen gene expression may be time-dependent with inhibition at the early time point, but stimulation later on.

Analysis of Gene Expression by Transfection of Regulatory DNA Sequences

Transfection studies may be performed to analyze the DNA sequences and transcription factors involved in the regulation of gene expression using plasmid vectors, in which the expression of reporter genes such as CAT or luciferaseis driven by gene regulatory sequences, such as those regulating type II collagen gene (COL2A1) transcription. Coexpression of wild-type or dominant-negative mutants of transcription factors, protein kinases, and other regulatory molecules, mediated by plasmid or adenoviral vectors may be performed to further dissect the mechanisms involved.

Transient transfections using luciferase reporter plasmids

1. Prepare plasmids using the EndoFree Plasmid Maxi Kit, according the manufacturer's instructions, to generate endotoxin-free DNA.
2. On the day before transfection, seed cells in six-well tissue-culture plates at 2.5–5 × 10^5 cells/well in DMEM/F-12 containing 10% FCS. Hyaluronidase may be added during these medium changes to facilitate the infection efficiency.
3. Change the medium to fresh growth medium 24 h and 3 h before the transfection to ensure that the cells are actively dividing. Prepare lipid/DNA complexes in serum-free DMEM/F-12 or Opti-MEM using LipofectAMINE+ or FuGENE 6, according to the

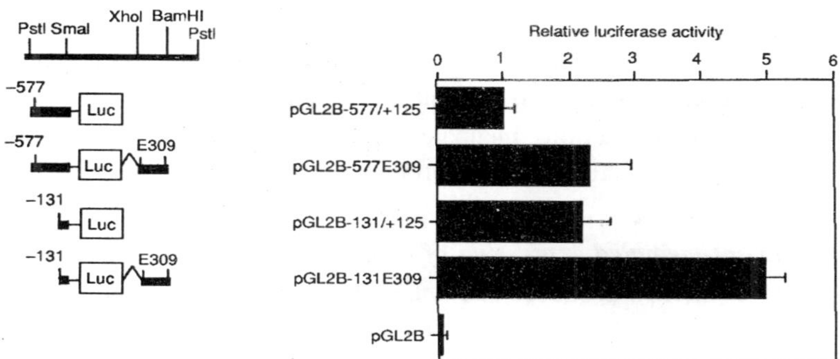

Fig. 9.2. Expression of the COL2A1 promoter human articular chondrocytes.

manufacturer's protocol. Prepare in bulk for multiple transfections. Do not vortex at any step:

(a) LipofectAMINE+: For each well, add 92 μL of serum-free medium to a small sterile polypropylene tube, add 1 μL of plasmid DNA, and tap gently to mix. Add 6 μL of PLUS reagent, mix and incubate for 15 min at room temperature. Dilute 4 μL of LipofectAMINE+ reagent into 100 μL of serum-free medium, mix and add to each reaction mixture. Mix and leave at room temperature for an additional 15–30 min at room temperature.

(b) FuGENE 6: For each well, add 96 μL of serum-free medium to a small sterile polypropylene tube, add 3 μL of FuGENE 6 reagent, and tap gently to mix. Add 1 μL plasmid of DNA (maximum of 1 μg) to the prediluted FuGENE 6 reagent and incubate for 15 min at room temperature.

4. While the lipid/DNA complexes are forming, replace culture medium on cells with serum-free medium to give a final volume of 1 mL. Add the lipid/DNA complex mixture dropwise to the well and incubate for 4 h at 37°C.
5. Dilute the transfection medium by adding to the wells an equal volume of DMEM/F-12 containing 2% Nutridoma-SP (or 20% FCS), and incubate 2 h to overnight. Add test agent without medium change and incubate further for 18 h (up to 48 h).

Cotransfections using plasmid vectors for expression of recombinant proteins

1. Prepare plasmids as described above.
2. Titrate each expression vector and its corresponding empty vector, at amounts ranging from 10 to 200 ng per well, against a fixed

amount of reporter vector. Equalize the total amount of reporter plus expression plasmid in each well (<1 μg/well) by adding empty vector and maintain equal volumes.

3. After cotransfection, incubate the cells for 18–24 h to permit expression of recombinant protein prior to treatment with test reagents.

Adenoviral-mediated expression of recombinant proteins

1. Infect the 293 producer cell line with adenoviral vector containing the cDNA encoding the wild-type or mutant protein to be coexpressed and determine the titer (moi) by standard techniques.
2. Incubate immortalized chondrocytes in DMEM/F-12 containing 10% FCS for 18 h following transfection of the reporter construct.
3. Remove medium and wash cells with PBS.
4. Add 1 mL of serum-free medium containing adenovirus at 1:125 moi. Incubate at 37°C for 90 min.
5. Add 1 mL of DMEM/F-12 containing 20% FCS and continue incubation for 18 h.
6. Change medium to fresh DMEM/F-12 containing 10% FCS or 1% Nutridoma-SP, incubate for 1 h, and treat with test agent for 18 h.

Luciferase assay

1. Prepare cell lysates by extraction with 200 μL of passive lysis buffer, which will passively lyse cells without the requirement of a freeze-thaw cycle. Scrape cells with policeman and transfer solubilized cells to 1.5-mL microcentrifuge tube. Microcentrifuge 5 min at maximum speed at 4°C, transfer supernatant to clean microcentrifuge tube, and store on ice.
2. Determine the protein content using the Coomassie Plus Protein Assay Reagent.
3. Determine luciferase activities using the Luciferase Assay System or equivalent, according to manufacturer's protocol. Mix, manually or automatically, 20 μL of cell lysate with 100 μL of Luciferase Assay Reagent and read in a luminometer. Normalize to the amount of protein (or internal control such as β-galactosidase) and express as relative activity against the empty vector or untreated control. Perform each treatment in triplicate wells and each experiment at least three times to ensure reproducibility and significance.

4. The Dual-Luciferase Reporter Assay, in which 20 ng of the pRL-TK Renilla luciferase control vector is included in the transfections, may be used routinely or as necessary to check the purity of new plasmid preparations or relative activities of mutant and wild-type constructs.

Notes

1. Batches of serum should be tested and selected on the basis of the capacity to support expression of chondrocyte-specific matrix gene expression. High capacity to induce cell proliferation is not necessarily associated with the ability to maintain phenotype.
2. Chondrocytes are quite resilient and tolerate the prolonged incubation times required for complete dissociation of the matrix and the absence of serum in the costal cartilage digestion. If the digestion is not complete by the end of the allotted time, then more collagenase solution may be added, or the suspension may be recovered and the fragments left behind for further digestion. These conditions result in suspensions that are essentially single cell, and therefore, it is not necessary to resort to filtration through a nylon mesh, as has been done by others when shorter digestion times are used. These considerations are important for decreasing the loss of chondrocytes during their isolation from valuable human cartilage specimens.
3. After initial plating of the primary cultures, the chondrocytes require 2–3 d before they have settled down and spread out completely. Culture for ~approx 4–7 d is required before reasonable amounts of total RNA may be extracted. Although the cultures may continue to express chondrocyte phenotype (e.g., type II collagen and aggrecan mRNAs) for several weeks, expression of nonspecific collagens I and III may begin as early as d 7 after isolation. Adult articular chondrocytes are strongly contact-inhibited and they may lose phenotype within 1–2 wk of monolayer culture. Juvenile costal chondrocytes continue to express chondrocyte phenotype (e.g., type II collagen mRNA) for several weeks and will form multilayer cultures. After they are subcultured, both types of chondrocytes cease the expression of chondrocyte matrix proteins, but this loss of phenotype is reversible and the cells may be redifferentiated in suspension culture within or on top of a nonadherent matrix.
4. Because chondrocytes adhere strongly to tissue culture plastic, possibly because of the presence of calcium ion-binding glycosaminoglycans in the pericellular matrix and cell membrane,

a trypsin-EDTA solution rather than trypsin alone should be used for full recovery of chondrocytes from tissue culture plastic during passaging. It is preferable not to use any antibiotics in order that any contamination that arises becomes apparent immediately. If necessary, standard concentrations of penicillin–streptomycin, gentamycin, and so on, that are suggested for fibroblast cultures are acceptable for use in chondrocyte cultures.

5. Primary chondrocyte cultures should be used for experimental analyses immediately before or just after confluence is reached to permit optimal matrix synthesis and cellular responsiveness. If the cells are not used or subcultured, they may be left at confluence for several weeks with weekly medium changes as long as the volume of the culture medium is maintained. If long-term culture results in the deposition of excessive matrix that is not easily digested with trypsin-EDTA, then a single-cell suspension may be obtained by using a dilute solution of collagenase (0.25%) and trypsin (0.25%) in PBS.
6. The synthetic activities of chondrocytes in monolayer culture are inversely related to proliferative activities. Thus, the expression of genes encoding matrix proteins and their deposition into the extracellular matrix increase compared to cell growth-associated genes. For experiments, the growth medium should not be changed within 3 d before addition of the test agent. Alternatively, the cells should be made quiescent by changing to serum-free medium supplemented with an insulin-containing serum substitute such as Nutridoma-SP or ITS+, followed 18 to 24 h later by the addition of the test agent of interest. Confluent cultures may tolerate serum-free medium containing 0.3% BSA for up to 48 h or longer.
7. Although the growth and maintenance of chondrocytes in primary culture or after subculture requires the use of 10% FCS, the loss of phenotype that occurs under these conditions may be delayed if the cells are plated at 4- to 10-fold higher density. Because high cell yields are not usually attainable from human cartilage sources, the reversibility of the loss of phenotype may be exploited by expanding the chondrocyte populations in monolayer cultures, redifferentiating the cells in fluid suspension culture and replating them in monolayer immediately before performing the experimental procedure.
8. After several passages in monolayer, chondrocytes may be redifferentiated by 2 wk or more of culture in alginate beads or in suspension over agarose or poly-HEMA.

9. The method for culture of chondrocytes in alginate beads has been adapted from previously published methods. For long-term alginate cultures, high viscosity alginate may provide more stable beads. Concentrations of serum as low as 0.5%, serum substitutes, or combinations of growth and differentiation factors or hormones have been used successfully, depending upon the experimental protocol, to permit chondrocyte phenotypic expression. Note that articular chondrocytes do not proliferate when cultured in fluid or gel suspension.
10. Ascorbate, which is required for synthesis and secretion of proteoglycans and collagens, is added daily to alginate or other 3D cultures to permit secretion and deposition of extracellular matrix, particularly when staining techniques are to be used. Add 25 μg/mL of ascorbate during the final 24–72 h of incubation when radiolabeling proteoglycans with ^{35}S-sulfate or collagens with ^{3}H-proline for characterization by SDS-PAGE. It is not necessary to maintain ascorbate in cultures if analysis of type II collagen or aggrecan mRNA is the endpoint, since effects on gene transcription may vary according to the time of incubation.
11. Culture of immortalized chondrocytes in 3D scaffolds is a useful approach for tissue engineering applications. The commercially available methods are recommended because of their ease of use. Published methods are available for fabricating collagen sponges and other 3D scaffolds, where the composition may be manipulated, for example, by using type II collagen and/or adding proteoglycans and other cartilage-specific matrix components. The biodegradable scaffolds are particularly useful if the cell-seeded scaffolds are to be implanted in animals. For studies entirely in vitro, where incubation periods of more than a few days are required, it is recommended that cultures be performed in wells that fit the size of the scaffolds. Otherwise, the culture surface of the well or dish should be coated with a nonadherent substrate or treated in such a way as to prevent attachment of cells that may migrate out from the sponges. Additional analytical methods have been described using Gelfoam and BD 3D scaffolds.
12. Biosynthetic labeling and immunocytochemistry procedures are readily performed on chondrocytes in a solid suspension system such as alginate, agarose, or collagen gels. Alginate culture may be the method of choice, since the chondrocytes are easily recovered by depolymerization of the alginate with a calcium chelator.

13. Various methods are available for analysis and characterization of proteoglycans. We have found the described methods to be a convenient and rapid approach for screening the relative amounts and molecular sizes of newly synthesized proteoglycans. Specific antibodies are available for more precise identification by either Western blotting or immunocytochemistry.
14. Although chondrocytes are generally less susceptible than monocyte/ macrophages and other immune cells to endotoxin, it is possible that the transfection conditions, the proliferative state of the cells, or other factors may sensitize the cells to low concentrations of endotoxin. Endotoxin itself induces and activates transcription factors that are common to inflammatory responses and may thus up- or downregulate the promoter of interest, thereby masking the response to a cytokine or growth factor.
15. Hyaluronidase added before and/or during transfections has been shown to increase transfection efficiencies in chondrocytes.
16. The total amount of plasmid to be transfected, including reporter, expression, and internal control plasmids, should not exceed 1 μg and the optimal amount for the culture system should be tested empirically. If variable amounts of expression vector, for example, are included, the total amount of plasmid in each well should be equalized by the addition of the empty vector. Wells transfected with appropriate empty vector controls, without and with treatment with test agent, should also be included.
17. The times of incubation following transfection before addition of the test agent may vary according to the cell density and culture condition and should be tested empirically. Nutridoma-SP or other serum substitute may be used for experiments that require quiescent cells for the reasons indicated. Note also that test agents should be added without medium change to avoid induction of pathways by serum growth factors or other medium constituents.

10

Apoptosis

Populations of cells which will divide forever in culture (*continuous cell lines*) differ from those which cease to expand after several weeks or months of growth (*finite cell lines*). The end of the proliferative lifespan of a cell culture has been loosely termed as either its '*senescence*' or its '*death*'. These states are distinct biochemical entities and the causes of any failure to thrive in a cell population may involve both processes. The purpose of this chapter is to describe methods for studying senescence, apoptosis, and necrosis.

Cellular Senescence

The term *senescence* is used in two distinct ways, applying either to individual cells or to entire cultures, and this can result in confusion. Senescence of an entire culture, sometimes called 'the *Phase III phenomenon*', is a failure of the culture to proliferate under conditions which had previously allowed sustained cell growth. This growth is generally measured as *population doublings* (PDs), calculated as the number of times the cell population doubles in number during the course of culture.

$$PD = \frac{\log 10\,(\text{number cells harvested}) - \log 10\,(\text{number cells seeded})}{\log 10^2}$$

Senescence has been studied mainly in normal human fibroblasts, cultures of which go through 30-60 population doublings, by which time virtually the entire culture is composed of senescent cells. These senescent fibroblasts are responsible for the decline in growth potential of ageing cell populations. The characteristic feature of senescent cells is their failure to divide in response to a mitotic stimulus. Senescent cells do not appear suddenly at Phase III, but instead compose a gradually increasing percentage of the population throughout the lifespan of the culture.

Senescence is distinct from quiescence, a transiently growth arrested state which can be induced by reducing the concentration of serum or through contact inhibition. *Quiescence* can be reversed by passaging the cells or by the re-addition of serum. Confusingly, both quiescence and senescence are referred to as the G_0 phase of the cell cycle (sometimes distinguished as G_0^Q or G_0^S respectively).

Cell Death

Necrosis can be triggered by simple physical trauma or gross damage. Programmed cell death or apoptosis is initiated by an individual cell in response to a specific stimulus or as a result of an inappropriate set of signals received from the external environment. There are several subtypes of apoptosis utilizing distinct pathways and patterns of gene expression. However, they converge on a common set of phenorypic markers and behaviours, thus allowing the fraction of apoptotic cells to be defined regardless of the pathway by which cell death is triggered.

Differentiation and De-differentiation

Primary keratinocytes in culture retain the ability to differentiate under the correct culture conditions. Similarly, cultured primary T cells retain the appropriate CD markers and the ability to kill target cells as measured by ^{31}Chromium release assays. The process of differentiation can also result in growth arrest and contribute to the failure of a culture to expand. The contribution of differentiation to lack of growth within a culture must be addressed on a cell type by cell type basis using known markers of the process (such as involucrin for keratinocytes), where available. Differentiation and senescence have been shown to be separate processes in some cell types.

Cell cultures can also appear to de-differentiate or lose the phenotype of the tissue from which they were derived. *In vitro* culture favours the relatively undifferentiated proliferating cells, which may not express tissue-specific markers either *in vivo* or *in vitro,* and this is the usual reason why primary cultures and the cell lines derived from them tend to progressively lose the differentiated phenotype.

In contrast to de-differentiation, de-adaptation is defined as the loss of the stimulus to make tissue-specific proteins rather than the capacity to do so. Detailed study may allow the missing stimulus to be identified and thus restore the specific phenotype.

Simple Measures of the Population Dynamics of Primary Cultures

The simplest static measure of growth, increase in cell number with time, provides no clues to the fraction of actively dividing, reproductively sterile, or dying cells within a given population. More complex static measures such as the determination of DNA content, RNA content, cellular ATP or protein concentration provide more information, but usually require more work.

The following techniques (used in combination on the same culture) allow the determination of the fraction of growing senescent, necrotic, and apoptotic cells and thus a complete picture of that culture's population dynamics. All depend on the culture of cells on 13 mm diameter Number 1 thickness coverslips cleaned according to the following protocol.

Protocol - 1. Cleaning of coverslips for cell culture

Equipment and reagents

1. 13 mm, Number 1 thickness glass coverslips (Merck)
2. Fine forceps
3. 100 mm diameter glass Petri dishes
4. 100 mm diameter filter papeı (Merck)
5. Standard laboratory autoclave
6. Benchtop heater
7. Absorbent paper
8. Flow 7X tissue culture grade detergent (ICN-Flow)
9. Double distilled water
10. Absolute ethanol

Method

1. Add 100 coverslips to 1 litre of a 5% (v/v) solution of Flow 7X in double distilled water. This will be cloudy.
2. Heat until the solution clears (usually 40-50°C). Maintain the coverslips at this temperature for a minimum of 1 h.
3. Wash the coverslips seven times in fresh doubled distilled water and once in 100 ml of absolute ethanol.
4. Shake the coverslips onto a sheet of absorbent paper. This absorbs the ethanol.
5. Separate the coverslips using fine forceps and place individually on a piece of filter paper within a glass Petri dish. Approx. 20 coverslips can be placed into each dish in this fashion.

6. Seal the Petri dish using autoclave tape and autoclave.
7. Open the autoclaved dish only under sterile conditions.

Measurement of the Growth Fraction

Normal somatic cells do not divide in synchrony during normal growth. Primary fibroblast cultures are composed of a mixture of clones with very variable growth potentials, This behaviour was shown not to be an inherent property of individual clones, because re-cloning of a clone with a long lifespan gave subclones with a range of division potentials. These cloning experiments demonstrated that the reproductive ability of fibroblasts was to some degree determined by a chance (*stochastic*) process, and that smaller clones predominated as the culture aged. Some evidence suggests that this process in human cells is due to the gradual attrition of telomeric sequences.

Work on bulk cultures complemented these single cell analyses. Cristofalo and Scharf carried out an analysis of the cell kinetics using embryonic fibroblasts, pulse labelling the cells with [^{3}H]thymidine at every passage throughout the lifespan of the cultures. The fraction of cells that entered S phase was then estimated by autoradiography. The assumption was made that an unlabelled cell is senescent. It was observed that unlabelled cells were present in cultures of embryonic cells and that a few labelled cells were present in even the latest passage cultures. It was also shown that the fraction of unlabelled cells increased smoothly with serial passage. Additional experiments excluded lengthening of the cell cycle as a property of senescence. Thus, primary cultures are mixtures of senescent and growing cells, the proportions of which alter as the cultures age. Such kinetic analyses can provide valuable data on the rates of senescence in both normal cell cultures derived from different tissues and can be performed in three ways:

1. Short pulse label experiments designed to detect a representative growth fraction for comparative studies.
2. Cristofalo-Scharf style long label experiments designed to detect the senescent fraction of the culture.
3. Direct detection of cell cycle-related proteins which may be used as proliferation markers. This last technique has the advantage that labelling time is not a factor.

Protocol - 2. Labelling using 5-bromo-2'-deoxyurldlne (BrdU)

Equipment and reagents

1. Epifluorescence microscope with filters suitable for detection of fluorescein isothiocyanate (FTTC) and 4-6- diaminophenylindole

2. Fine forceps
3. Parafilm
4. 100 mm diameter glass Petri dishes
5. 100 mm diameter filter paper (Merck)
6. Glass slides
7. Vectashield mountant containing DAPI
8. Labelling reagent: BrdU, 3 mg/ml; 5-fluoro- 2'-deoxyuridine, 0.3 mg/ml
9. Mouse monoclonal antibody (IgG) to BrdU pre-diluted with DNase I
10. PolyctonaJ rabbit antiserum to mouse IgG conjugated with HTC
11. Cell type of interest cultured to subconfluence on coverslips
12. Phosphate-buffered saline (prepared using Oxoid PBS tablets)
13. PBS-FCS: 156 (v/v) FCS diluted in PBS
14. 95%-5% (v/v) ethanol-acetic acid for fixation

Method

A. Labelling of cells

1. Add labelling reagent at a dilution of 1:1000 to the culture medium. (Do not use medium which contain thymidine, such as Ham's F12, since this results in poor uptake of BrdU.)
2. Allow to pulse label for a suitable period. The labelling time of choice is dependent on the style of assay to be undertaken. Experiments designed to detect a representative growth fraction for comparative studies require a labelling time less than the S phase duration (we use 1 h labels). Experiments designed to detect the senescent fraction of the culture require a labelling time of 72 h or more (less than 5% label-incorporating nuclei can be taken to indicate culture senescence). However, the absolute growth fraction will be inaccurate since some cells will have divided during the labelling period. Intermediate labelling times (i,e. 24 h) are quite difficult to interpret and should be avoided.
3. Aspirate the medium. Quickly wash the cells in three changes of PBS.
4. Aspirate the last change of PBS. Flood the coverslip with ethanol-acetic acid. Fix at –20°C for 20 min.
5. Wash a further three times in PBS. Leave the third wash in contact with the coverslip.

B. Detection of BrdU incorporation

1. Prepare a '*humidified chamber*'. This consists of a glass Petri dish containing a water saturated filter paper on which is placed a square of Parafilm. This prevents desiccation during the assay.
2. Place 50 μl of anti-BrdU on the Parafilm, then using fine forceps gently place the coverslip face down on the drop of antibody (this ensures even spreading).
3. Incubate for a minimum of 90 min at room temperature (never at 4°C since DNase action is required to allow the antibody access to the DNA).
4. Wash three times in PBS (by dipping the coverslip at least ten times into each of a series of three Universal tubes).
5. Repeat step 2 substituting a 1:20 dilution of anti-mouse IgG in PBS-FCS.
6. Incubate overnight at 4°C.

C. Quantification of label-incorporating celts

1. Wash the coverslip three times in PBS as above, in addition include a final wash in double distilled water. This removes salts from the coverslip.
2. Place a drop of Vectashield mountant onto a clean glass slide and place the coverslip face down onto the mountant. The slide may be sealed with clear nail varnish if required.
3. View under epifluorescence using the DAPI filter to focus on the cell nuclei to prevent quenching (the DAPI signal may be low in intensity due to nuclease activity). Count the total number of nuclei within the field of view.
4. View the same field under FITC filters. A distinct punctate staining reveals incorporation of BrdU. Count the number of positive cells within the field of view.
5. Repeat steps 3 and 4 until the count has reached 400 positive or 1000 total nuclei, whichever comes first. The percentage of labelled to unlabelled cells is the 95% confidence limit of the growth fraction (corrected for labelling time).

Protocol - 3. Detection of total growth fraction using anti-sera to pKa

Equipment and reagents

1. Epifluorescence microscope with filters suitabie for detection of fluorescein isothiocyanate (FITC) and 4-6-diaminophenylindole (DAPI)

2. Fine forceps
3. Parafilm
4. 100 mm diameter glass Petri dishes
5. 100 mm diameter filter paper
6. Glass slides
7. Vectashield moimtant containing DAPI Mouse monoclonal antibody directed against human Ki-67. Polyclonal rabbit antiserum to mouse IgG conjugated with FITC. Cell type of interest cultured to subconfluence on coverslips PBS PBS-FCS Methanol/acetone (1:1) for fixation.

Method

1. Aspirate the medium. Quickly wash the cells in three changes of PBS.
2. Aspirate the last change of PBS. Flood the coverslip with methanol/acetone. Fix at 4°C for 5 min.
3. Wash a further three times in PBS. Leave the third wash in contact with the coverslip.
4. Perform irnmunocytochemical staining as described above, substituting 15 n.1 of anti-Ki-67 for anti-BrdU.
5. Visualize the Ki-67 positive fraction of the culture as described above. Ki-67 positive cells show a highly distinctive punctate nuclear staining under F1TC excitation. When a culture is moving from G0q, Ki-67 staining first becomes detectable in late G1 However it remains present throughout the cell cycle for cultures maintained in log phase growth.

Determination of the Necrotic Fraction of the Population

Several techniques are available for this purpose, of which the simplest remains trypan blue staining. This dye exclusion assay can be used either as part of a routine passage protocol or as a simple washing step on coversiips of unfixed material. However, tiypan blue exclusion provides only a rough estimate of viability and needs independent validation for the particular application before being used routinely.

A visually pleasing and highly discriminatory alternative is afforded by calcein AM-ethidium bromide staining, which we nave employed to estimate the fraction of living and necrotic cells both on coversiips and in collagen matrices, Calcein AM is membrane permeant and is cleaved by active esterases to yield a green cytoplasmic fluorescence whilst ethidium is membrane impermeant and labels only the DNA of cells with a compromised outer membrane. Unfixed material is required in both cases.

Protocol - 4. Determination of viable and necrotic fractions using calcein AM-ethidlum bromide staining

Equipment and reagents

1. Epifhiorescence microscope with filters suitable for detection of fluorescein isothiocyanate (FITC) and tetramethyl rhodamine isothiocyanate (TRITC)
2. Adherent live cells growing on coverslips prepared according to Protocol 1 Calcein AM fluorescent esterase substrate Ethidium homodimer 1 Phosphate-buffered saline (prepared using Oxoid PBS tablets)

Method

1. Wash the cells three times in PBS.
2. Add 0.05 mg/ml each of calcein AM and ethidium homodimer in PBS.
3. Incubate for a minimum of 10 min at 37°C.
4. View under a fluorescence microscope. Live cells show strong cytoplasmic fluorescence under FTTC excitation. Dead cells show nuclear fluorescence under TRITC filters.
5. Score 200 cells first under FITC and then TRITC filters. Express viability as a percentage.

Determination of the Senescent Fraction of the Population

Until recently there was no simple way to directly demonstrate that a single cell was senescent other than a combination of lack of label incorporation and metabolic viability. However a variation of the standard catalytic histochemical staining technique for β-galactosidase activity has been shown to discriminate between senescent cells and their growing counterparts, The mechanistic basis of this 'senescence-associated β-galactosidase' (SA-β) assay is unknown and may be related to the overexpression of the enzyme at senescence or to a simple increase in cell size correlating with entry into the permanently nondividing state. It may be that the majority of lysosomal enzymes are amenable to histochemical visualization by related techniques.

A series of cell types on which the assay has been validated has been published. Variants of the technique have also been employed. Although we do not routinely undertake dual labelling studies, SA-β staining is compatible with [^{3}H]thymidine pre-labetling (10 mCi/ml) and we have had success combining SA-β with BrdU labelling (72 hours) and antisera to Ki-67 (rabbit polyclonal). SA-β staining also supports cell permeabilization with 0,2% Triton X-100.

Protocol - 5. Demonstration of senescent cells by senescence-associated β-gatactosldase (SA-β) activity

Equipment and reagents

1. Light microscope
2. Glass slides
3. Other equipment for preparing glass coverslips as given in previous protocols
4. 0.1 M citric acid solution: 2.1 g citric acid monohydrate in 100 ml distilled water
5. Sodium phosphate solution: 2.84 g/100 ml sodium dibasic phosphate or 3.56 g/100 ml sodium dibasic phosphate dihydrate
6. Citric acid/sodium phosphate buffer (100 ml): 36.85 ml of 0.1 M citric acid solution and 63.15 ml of 0.2 M sodium phosphate (dibasic) solution: verify that the pH is 6.0
7. X-gal powder; stored in dark at -20°C
8. 5 M sodium chloride
9. Dimethylfbrmamide (DMF)
10. Potassium ferrocyanide
11. Potassium ferricyanide
12. 1 M magnesium chloride
13. Apathy's syrup, histamount, or similar aqueous mounting medium
14. Fixative: 2% paraformaldehyde/0.2% glutaraldehyde in PBS; SA-β staining is also compatible with 3% formaldehyde.

Method

A. Preparation of staining solution

1. For 20 ml staining solution combine 4 ml citric acid/sodium phosphate buffer with 20 mg of X-gal (in a few drops of DMF only since this inhibits the β-galactosidase). Then add:
   (a) 1 ml of 100 mM potassium ferrocyanide
   (b) 1 ml of 100 mM potassium ferricyanide
   (c) 0.6 ml of 5 M sodium chloride
   (d) 40 μL of 1 M magnesium chloride
   (e) 13.4 ml double distilled water

B. Detection of SA-β activity in adherent cells

1. Wash material twice in PBS.
2. Fix for 3-5 min at room temperature. Prolonged fixation will inactivate the enzyme.
3. Repeat washes in PBS.
4. Add staining solution (1-2 ml per 35 mm diameter dish).
5. Incubate at 37°C.
6. Senescent cells display a dense blue colour. This is detectable in some cells within 2 h, staining is usually maximal after 12-16 h.
7. Material can be mounted in aqueous mountants (e.g. Apathy's syrup) and can be counterstained with fast green or similar colour compatible stain.

Determination of the Apoptotic Fraction of the Population Using TUNEL

Activation of endogenous nuclease activity is considered to be a key biochemical event in apoptosis, leading to the cleavage of DNA into nucleosome-sized fragments. This process gives rise to free 3' hydroxyl groups which can be extended using terminal transferase or DNA polymerase. If labelled nucleotides are included in the end-labelling mixture, apoptotic nuclei can be directly visualized. This TdT-mediated dUTP nick end-labelling assay (TUNEL) is fast and effective but should not be employed on any coverslips on which the cells have been pre-pulsed with BrdU since this undergoes photolysis in daylight. This can result in a large number of free 3' hydroxyl groups within the DNA of any labelled cells and thus a falsely high TUNEL reading.

Protocol - 6. Demonstration of apoptotic cells by TdT-mediated dUTP nick end-labelling (TUNEL)

Equipment and reagents

1. Epifluorescence microscope with filters suitable for detection of fluorescein isothiocyanate (FTTC) and 4-6-diaminophenylindole (DAPI)
2. Vectashield mountant containing DAPI
3. Glass slides
4. Other equipment for handling glass coverslips as given in previous protocols Phosphate-buffered saline (prepared using Oxoid PBS tablets) (Oxoid) 3% (w/v) paraformaldehyde in PBS for fixationa 0.1% (v/v) Triton X-100 in PBS Pre-diluted terminal transferase

solution and nucleotide labelling solution containing fluorescein-conjugated dUTP.

Method

1. Aspirate the medium and wash the cells as described above.
2. Aspirate the last change of PBS. Flood the coverslip with the paraformaldehyde solution. Incubate at room temperature for 10 min to fix. Wash with PBS as described above.
3. Add 0.1% Triton X-100 and incubate for 5 min at room temperature, this permeabilizes the cells and allows the terminal transferase access to the nuclear DNA. Wash a further three times with PBS.
4. Add 50 μl of terminal transferase solution to 450 μl of labelling solution to reconstitute a working TUNEL solution.
5. Spot 50 ul of TUNEL solution onto Parafilm in a humidified chamber and incubate the coverslip with the solution as described above.
6. Incubate the humidified chamber for 90 min at room temperature then wash, mount, and quantify as described above. Apoptotic nuclei show strong but diffuse FITC fluorescence.

Other Techniques for Analysing Population Dynamics

Growth and apoptosis require ATP. Apoptosis, but not growth arrest, is accompanied by an elevation in ADP levels. Therefore measuring the relative levels of ATP and ADP in cultures, and calculating the ADP/ATP ratio, can be informative. Necrosis is characterized by a rapid decline in ATP levels, so it may be possible to discriminate between the forms of cell death. A novel assay system has been developed in kit form which is based upon estimation of ATP/ADP levels by luminometry. The assay is performed in 96-well plates, allowing high throughput.

The simplest method for detecting dead or dying cells within a population is by microscopic observation. Phase-contrast will reveal floating or dying cells as rounded, dense bodies. Fluorescence microscopy allows for staining of cultures with dyes such as DAPI or Hoechst 33258. Cells undergoing apoptosis have fragmented or irregularly stained chromatin whereas unaffected cells display relatively homogeneous staining.

Detection of Apoptosis using DNA Laddering

Some apoptotic pathways cleave genomic DNA into mono- and oligonucleosomal DNA fragments. Visualization of this process following agarose gel electrophoresis of DNA reveals multiples of the 180-200 bp nucleosomal fragment in a characteristic '*ladder*' pattern. The occurrence of DNA fragmentation should not be considered as definitive for cells undergoing apoptosis, since some cells undergoing necrosis also display '*ladders*'. Conversely, several cell types do not show ladders despite undergoing apoptosis. Thus, detection of apoptosis by DNA laddering should be qualified by other tests.

Protocol - 7. Analysis of DNA fragmentation

Reagents

1. Buffer 1:10 mM EDTA, 0.5% (w/v) sodium lauryl sarkosinate, 50 mM Tris-HCI pH 8.0
2. Proteinase K: 20 mg/ml stocks stored at –20°C; prepare a fresh working solution (5 μl of stock plus 195 μl of buffer 1) for each assay and keep on ice before use
3. RNase A: 10 mg/ml stock in water, boiled for 10 min to destroy DNases; dilute 10 μl of stock with 190 μl of buffer 1 for use
4. Ethidium bromide: 10 mg/ml stock in TE TPE buffer: 342 g of Tris base added to 1 litre of dH_2O, add 46.5 ml phosphoric acid, 240 ml of a 0.25 M solution of EDTA pH 8.0, mix to dissolve, then make up to 3 litres Loading buffer: 10 mM EDTA pH 8.0. 1% (w/v) low melting point agarose, 40% (w/v) sucrose, heat to dissolve, aliquot into 0.5 ml

Method

1. Transfer 5 × 10^5 to 10^6 cells to a 1.5 ml microcerttrifuge tube and pellet by centrifugation at 4000 r.p.m, for 5 min at 4°C.
2. Remove supernatant and recentrifuge for 15 sec. Remove the last of the supernatant.
3. Resuspend pellet by vortexing in 20 μl of buffer 1.b
4. Incubate at 50°C for 1 h in a dry block.
5. Spin down condensate from lid and add 10 μl of RNase solution. Vortex and incubate for a further hour at 50°C.
6. Meanwhile, make up a 2% (w/v) agarose solution in TPE buffer. Heat in a microwave to melt the agarose. Cool and add 1 μl ethidium bromide for 100 ml gel solution.
7. Whilst agarose is cooling, place a well comb in a prepared gel tray about 1 cm from one end. Pour in the agarose solution ensuring no bubbles form in the gel. Allow to set on a level surface for at least 20 min.

8. Centrifuge cell suspension for 15 sec to collect condensate from the lid of the tube. Add 10 μl of loading buffer which has previously been incubated at 70°C to melt the agarose.
9. Mix by vortexing, centrifuge briefly, and reheat to 70°C.
10. Load samplesd into dry wellsc of the agarose gel using pipette tips just narrower than the well.
11. Carefully flood the gel tank with TPE buffer and run at 50 V for 1.5 h.
12. Visualize DNA using a UV light source.
    (a) Cell pellet can be stored frozen at -20°C for up to two weeks.
    (b) EDTA chelates magnesium and calcium ions thus reducing further nuclease activity. Sodium lauryl sarkosinate denatures histones thereby releasing DNA. Proteinase K degrades proteins including DNases and histones.
    (c) Low power (70%) for 3-4 min. Do not boil and avoid bubbles.
    (d) Include a 100 bp ladder for reference in one well.
    (e) If wells are flooded prior to loading, high molecular weight DNA can float out and cause spurious banding patterns.

Inhibition of Apoptosis Using Peptide Inhibitors of Caspases

A critical step in the pathway to programmed cell death involves the activation of a series of proteases known collectively as caspases (cysteinyl aspartatespecific proteinases). The prototypic caspasc, caspase 1, was identified as the protease responsible for cleaving interleukin 1β and was known as ICE (interleukin 1β converting enzyme). The caspases cleave substrates to render them inactive or to activate precursor forms. In this way caspases initiate a cascading system of proteases leading to cell death. Thus, inhibition of caspase activities, particularly at the apex of a cascade, will prevent apoptosis and provide strong evidence for apoptosis. Inhibition of caspases is possible using specific peptide inhibitors. These inhibitors can be either caspase-spedfic or general. The protocol below describes an inhibition assay using a general inhibitor of all caspases. A range of other peptide inhibitors can then be used to assess the involvement of specific caspases in the apoptotic pathway under examination.

Protocol - 8. Inhibition of caspase activity by peptide inhibitors

Reagents

1. 50 mM stock of peptide inhibitor in DMSO
2. 50 mM stock of negative control peptide, CBZ-Phe-Ala.fmk

Method

1. Pre-incubate cells for 1 h in the presence of inhibitor or negative control peptide.
2. Remove medium from culture dish and treat with suspected apoptosis-inducing agent.
3. Replace medium containing inhibitor/control peptide.
4. Maintain cultures for appropriate time to assess occurrence of apoptosis. a For most cellular assays the optimum concentration of each inhibitor required to block apoptosis varies from 50-200 u*M* dependent upon the cell type used.

Determination of the Non-dividing Fraction of a Population by Simplified Haptotactic Assays, 'Ponten Plates'

Haptotactic island analysis is an elegant, definitive, and non-invasive, but little used, assay for the presence of senescent cells (and phenotypic heterogeneity in general) within a bulk culture. The essence of the method is to '*trap*' single live cells on colonizable islands of palladium shadowed in a pattern onto a nonadhesive substrate. A larger area of palladium metal is then shadowed around the islands. First developed by Carter the method was significantly improved first by Westermark and then by Ponten and co-workers. Its advantage is that it allows the behaviour of single cells to be studied within a bulk culture without the potential pitfalls which accompany the changes of density required in a standard cloning experiment. The method described below is the original Westermark technique which, unlike that of Ponten, does not require custom-fabricated photolithographic masks but instead uses a standard copper electron microscope grid. The disadvantage of Westermark's technique is that several cell types (including many cell lines) are able to bridge the gaps between the islands because they arc rather close together. We have used the technique with human glia and fibroblasts.

Protocol - 9. Simplified haptotactic analysis

Equipment and reagents

1. 35 mm standard tissue culture dishes.
2. 200 mesh copper electron microscope grids.
3. Cloning rings Palladium metal and shadowing equipment Molecular biology grade agarose.

Method

1. Pour a boiling 1% (w/v) solution of agarose into each Petri dish.

2. Immediately suck off the surplus solution to leave a thin film behind. Leave to dry at 37°C for 1-2 h.
3. Add a standard copper EM grid (or a grid-pattern mask produced by photoetching) to the agarose-coated bottom of the Petri dish.
4. Evaporate palladium (or gold) onto the surface at 10-4 torr using a conventional EM evaporator. This produces a grid pattern and a thick metal 'ring'. The palladium islands should have distinct borders.
5. Place a cloning ring over the area of grids. Seed cells at cloning density (such that an average of one cell will attach to any one island) within the ring and at normal density outside it. This allows the cells on the islands to cross-feed and function at normal density.
6. View cell attachment after 24 h. Remove old medium (together with any floating cells). Note the number of '*one-cell*', '*two-cell*', and '*three-cell islands*'.
7. Allow the culture to proliferate for seven to ten days. Note the cell number on each island daily. Senescent cells will not divide and will remain as one-cell colonies. *Note*: The palladium islands can be manipulated by attaching serum proteins, polylysme, and other attachment factors to enhance cell adhesion.

Determination of Telomerase Activity and Telomere Length

Strong evidence now exists that the loss of telomeric DNA that accompanies DNA replication in several human cell types is the primary means by which such cells count divisions. When telomerase, an enzyme that maintains telomere length, is expressed at sufficient levels, the onset of senescence in these cells is also blocked. The simplest interpretation is that, in some cells, senescence is '*telomere driven*'. However recent evidence suggests that telomeredriven senescence is not found in every human cell type and does not appear to exist at all in some species. In Syrian hamster fibroblasts, senescence occurs in the presence of telomerase. The following assays estimate telomerase activity by the telomerase repeat amplification or TRAP assay and establish whether telomere length is stably maintained by terminal restriction fragment or TRF analysis.

A TRAP assay consists of three stages:

1. The production of a cell extract potentially containing telomerase.
2. Extension of a target primer by the telomerase in the extract (telomerase adds the repeat sequence TTAGGG).

3. A polymerase chain reaction step which amplifies the extended products to detectable levels.

Terminal restriction fragment (TRF) length analysis detects the length of chromosomal DNA containing the telomere and undigested subtelomeric DNA. Genomic DNA is digested with a combination of restriction enzymes that cleave frequently, but do not cut within the $(TTAGGG)_n$ arrays at the telomeres, followed by agarose gel electrophoresis, Southern blotting, and finally detection of the $(TTAGGG)_n$-containing terminal restriction fragments with an appropriate hybridization probe.

The main drawback of TRF analysis is the large number (10^5 to 10^6) of cells that are required. This is not a protocol that can be applied to single cells or an in situ format. TRF gives the mean telomere length of a population of cells and provides no information about the distributions of telomere length within the population or within the chromosomes of an individual cell. For these measurements, more sophisticated methods such as *peptide nucleic acid* (PNA) probes for telomere in situ hybridization coupled to extremely sensitive imaging equipment must be used. Other TRF protocols have been developed to use in-gel hybridization (which avoids technical problems related to Southern transfer of large fragments). The protocol below is of this type and is reliable.

Protocol - 10. Demonstration of telomerase activity by TRAP

Equipment and reagents

1. All reagents are from Sigma unless otherwise stated.
2. RNase-free Gilson pipette tips (Sarstedt).
3. DEPC treated microcentrifuge tubes
4. DEPC treated polycarbonate centrifuge tubes (Beckman)
5. Beckman benchtop ultracentrifuge with TLA 100.2 rotor at 4°C
6. PCR machine
7. Growing cells from which to obtain $100 extract; at least 10^6 cells are required
9. DEPC treated PBS
10. Trypsin/EDTA solution

Method

A. Preparation of $100 cell extract

The key to this procedure is the production of the cell extract in a RNase-free environment. *SWO* refers to the fact that the extract consists of the cellular supernatant following a 100000 g centrifugation step.

1. Detect adherent cells by trypsin dispersion.
2. Wash once in standard cell culture medium.
3. Wash twice in 10 ml ice-cold DEPC treated PBS.
4. Pellet the cells by centrifugation at 350 g for 5 min. This minimizes the chances of early lysis.
5. Resuspend the cells in 1 ml of wash buffer. Transfer the cell suspension to an Gppendorf tube and spin at 15 000 g for 2 min at 4°C. Resuspend the cells in 18.5 μl of lysis buffer per 10^6 cells.
6. Incubate the samples on ice for 30 min. The length of this incubation is crucial.
7. Load the samples into DEPC treated polycarbonate centrifuge tubes immediately after the end of the 30 min.
8. Immediately centrifuge the samples at 100 000 g for 30 min at 4°C. This equates to 53000 r.p.m. on a TLA 100, 2 rotor.
9. Following centrifugation remove the supernatant and snap-freeze on dry ice in 10 μl aliquots in DEPC treated microtubes.
10. Estimate the protein concentration of a sample of supernatant (by Bradford assay). This should be approx. 5-10 mg/ml.

B. TS primer extension (RNase-free conditions should be employed throughout!)

In order to standardize this part of the assay and reduce the effect of pipetting errors, a '*master mix*' or '*pre-mix*' is made corresponding to the number of reactions plus two more, A four reaction mix is summarized below.

1. Combine 100 ul of 2 × TRAP assay reaction buffer with 4 μg T4 gene 32 protein.
2. Add 400 ng TS primer.
3. Add 4 μl $100 extract,
4. Incubate under oil in a PCR machine at 30°C for 30 min.
5. Heat to 92°C to destroy the telomerase.

PCR amplification of extension products Again a '*pre-mix*' equal to the number of reactions plus two is made to limit dilution errors.

1. Add 100 μg of the CX primer per reaction.
2. Add 2.5 U of Toq polymerase per reaction.
3. Subject the sample to 31 cycles of denaturation (92°C, 30 sec), annealing (50°C, 30 sec), and extension (72°C, 90 sec).

4. Resolve the PCR products on a 10% polyacrylamide gel containing bromophenol blue. Run the gel at about 300 V until the bromophenol blue is approx, 3 cm from the bottom (this takes about 4 h).
5. Stain the gel using 1:10 000 dilution Sybr gold and view under blue fluorescence on a STORM system. An extract containing telomerase will produce a DNA ladder with 6 bp periodicity.

An alternative strategy which may be more applicable to some laboratories is to end-label the TS primer with [γ-^{32}P]ATP using T4 polynucleotide kinase. The resulting gel can then be dried and exposed to films or phosphor screens.

Protocol - 11. Determination of terminal restriction fragment (TRF) length by in-gel hybridization

Equipment and reagents

1. Standard gel electrophoresis equipment (power pack, tanks, combs)
2. Gel dryer
3. Hybridization oven and bottles (Hybaid)
4. Genomic DNA (at least 1 μg) from test cells of interest
5. Hinfl and Rsal restriction enzymes (Amersham)
6. TBE
7. 10 x buffer M: 100 mM Tris-HCl pH 7.5, 100 mM $MgCl_2$, 10 mM DTT, 500 mM NaCl
8. ^{32}P-labelled HindIII markers Denaturing buffer: 1.5MNaCl, 0.5M NaOH (to prepare add 87.66 g/litre NaCl and 20 g/litre NaOH in double distilled water) Neutralizing buffer: 1.5 M NaCl, 0,5 M Tris pH 8.0 (to prepare add 87.66 g/litre NaCl and 60.57 g/litre Tris base; adjust to pH 8.0 with HC1) Hybridization solution: 5 $\times$ SSC, 0.1 $\times$ phosphate wash (1 X contains 5 mM sodium pyrophosphate, 100 mM Na_2HPO_4) 5 $\times$ Denhardt's solution Tel 2 probe: 5'-CCCTAACCCTAACCCTAA-3'.

Method

1. Mix the following:
   (a) 10 μl (100 (100 uCi[^{32}P]ATP (3000 Ci/mmol)
   (b) 2 μl of 10 $\times$ T4 polynucleotide kinase buffer
   (c) 1 μl of T4 polynucleotide kinase
   (d) 500 μg of unlabelled Tel 2
   (e) Water to 20 μl

2. Incubate for 1 h at 37°C, then heat at 90°C for 5 min,
3. Separate Tel 2 from the unincorporated label on a 2 ml volume P4 gel column. Collect 0.5 ml fractions. The first fraction is the column void volume. The second fraction contains the probe.
4. Alternatively the probe can be purified by centrifugation on Quick Spin colum.

In-gel hybridization

1. Digest 1 ug of genomic DNA in 30 μl of buffer M containing 15 U of *Hinfi* and *Rsal*.
2. Run out 1 μl of sample on a 0.7% agarose gel to test if the DNA is fully digested.
3. If the digest has been successful then load the sample on a 15 × 15 cm 0.5% agarose gel (about 200 ml) in TBE. One lane should contain ^{32}P-labelled HindIII markers. Load all the samples in 6 × Ficoll loading buffer containing xylene cyanol and bromophenol blue. The samples should be run in at 100 V for 20 min followed by 20 V overnight.
4. When the bromophenol blue has run approx. three-quarters of the way down the gel remove it from the tank.
5. Denature the DNA by washing with denaturing buffer for 15 min.
6. Wash the gel in neutralizing buffer for 10 min.
7. Dry the gel at room temperature for 1 h (i.e. under vacuum but with the lid of gel drier up) followed by a further 30 min drying at 50°C.
8. In order to remove the gel it is necessary to wet the paper with 5 × SSC, Gently lift the agarose gel off the paper. At this point the gel can be stored at -80°C wrapped in Saran, This method does not require a pre-hybridization step.
9. Hybridize the gel at 37°C overnight in 25 ml of hybridization solution (large Hybaid bottles are useful for this) with 500 ng of Tel 2 probe.
10. Wash the gel as below:
    (a) Three times for 7 min at 37°C in bottles with 0.1 × SSC.
    (b) Twice for 7 min each at RT in bottles with 0.1 × SSC.
    (c) Once for 7 min in tray with 0.1 × SSC pre-heated to 37°C.
11. Wrap the gel in Saran and place in a cassette with autoradiographic film. If high background is a problem further washes can be performed. When the gel is not in use it should be kept at -80°C as the signal tends to become more diffuse with time.

The DNA appears as a smear, representing the telomeres from all the chromosomes from a population of cells. The smear demonstrates the heterogeneous nature of telomere length within and between cells. Cells with shorter telomeres exhibit a faint low molecular weight smear while cells with long telomeres show a more intense higher molecular weight signal. The use of densitometry can assign a quantitative value to the mean terminal restriction fragment length. It should be kept in mind that the TRF is not formally equivalent to the telomere as the fragments contain subtelomeric sequences of 1-2 kb that are resistant to digestion.

INDEX

K

L

M

N

O

P

Q